Selecting and Implementing an Integrated Library System

Chandos
Information Professional Series
Series Editor: Ruth Rikowski
(email: Rikowskigr@aol.com)

Chandos' new series of books is aimed at the busy information professional. They have been specially commissioned to provide the reader with an authoritative view of current thinking. They are designed to provide easy-to-read and (most importantly) practical coverage of topics that are of interest to librarians and other information professionals. If you would like a full listing of current and forthcoming titles, please visit www.chandospublishing.com.

New authors: we are always pleased to receive ideas for new titles; if you would like to write a book for Chandos, please contact Dr Glyn Jones on g.jones.2@elsevier.com or telephone + 44 (0) 1865 843000.

Selecting and Implementing an Integrated Library System

The Most Important Decision You Will Ever Make

Richard M. Jost

ELSEVIER

AMSTERDAM • BOSTON • HEIDELBERG • LONDON
NEW YORK • OXFORD • PARIS • SAN DIEGO
SAN FRANCISCO • SINGAPORE • SYDNEY • TOKYO

Chandos Publishing is an imprint of Elsevier

CP CHANDOS PUBLISHING

Chandos Publishing is an imprint of Elsevier
225 Wyman Street, Waltham, MA 02451, USA
Langford Lane, Kidlington, OX5 1GB, UK

ISBN: 978-0-08-100153-0

British Library Cataloguing-in-Publication Data
A catalogue record for this book is available from the British Library

Library of Congress Cataloging-in-Publication Data
A catalog record for this book is available from the Library of Congress

For information on all Chandos Publishing
visit our website at http://store.elsevier.com/

Dedication

To Charlie, of course.

Contents

About the Author

Choosing an Integrated Library System: The most important decision you will ever make by Richard M. Jost.

Richard Jost is currently the Information Systems Coordinator at the University of Washington Marian Gould Gallagher Law Library in Seattle. Previously, he served as the Assistant Librarian for Technical Services at the Gallagher Law Library from 1992 to 2008 and Assistant Librarian for Technical Services at the University of Colorado School of Law Library from 1988 to 1991. He has taught a class on library technology systems at the University of Washington Information School since 2000 and has been active in local, regional, and national law library associations. He holds a Master's in Library and Information Science from the University of Denver (1985), a Master of Arts in International Studies from the American University (1980), and a Bachelor of Arts in History from the State University of New York (SUNY) College at Cortland (1976).

List of Figures

Introduction

"Technology is easy, people are hard." This is the first thing that I tell my students as I begin every new class on library technology at the University of Washington Information School. As you can imagine, I get skeptical looks from some of the students as they enrolled in my class because they believe that library technology is the difficult part of that equation. And from other students, I get the distinct impression that the reason they wanted to take my class was so that they did not have to think about people at all – wasn't that the whole point of working in the technology field?

What I am trying to make them aware of is the importance of the human factor in any technology endeavor. The management aspects of a library project are equal to or even more important than the technology part. You can purchase the most technologically advanced library system on the marketplace, but without the "people skills" to have it embraced by the library staff, it is worthless. Technology will not solve any problems unless you can get the staff to use it effectively.

The modern systems librarian, who is often the staff person in charge of the selection and implementation of an integrated library system, has one of the most challenging positions in the library. This person needs to have excellent technology skills to assess the proposed systems that the library might acquire but also the skills to negotiate contracts, train staff members, and to work with the library administration on planning and budgets. In addition, he or she will also have to be familiar with change management, project management, and how to be successful in communicating with a wide variety of stakeholders.

The selection and implementation of an integrated library system is one of the most important decisions that a library can make. The choice of a new system will provide the library with a blueprint for its future operations, both for the staff and its patron community. Not only does it represent a significant cost in terms of funds, but there is also the consideration of staff costs – training, productivity levels, stress – that can have a profound effect on library operations. In addition, there is also the cost of migrating data between systems and whether that will be done effortlessly or result in a major cleanup project at the end of the migration. For all these reasons (and many more) the selection of a library system must be undertaken with the knowledge and skills to do it correctly. Hopefully, this book will provide some guidance to those library staff members selecting their very first system as well as to those veterans who might be migrating to a next-generation system.

Brief History of Library Technology

1

Brief history of library automation

From the very beginning of libraries, the control of collections has been the main goal of librarians. All the way back to antiquity, the ancient library of Alexandria maintained a listing of the papyrus rolls that it held, adding details to each annotation to form a unique description (Lerner, 2009, p. 16). In the Middle Ages, this practice of printed lists continued with the lists serving more as an inventory for these libraries, often housed in monasteries and that were only intended for the sole use of the monks (Lerner, 2009, p. 33). Lending of books between religious groups was done on a very limited basis (mostly for the purpose of copying) with no public access provided to these collections (Lerner, 2009, pp. 34–35).

As collections grew in size and scope, so too did the number of institutions that were maintaining their own library collections. The growth of cities and universities in the Renaissance spurred this increase as well as the number of wealthy private individuals (including royal courts) who were building their own collections. Palaces such as Versailles in France and the Winter Palace in Russia had magnificent libraries and extensive collections that often served as the basis for future national libraries.

Classification

Needing to provide better access to this growing number of books on diverse subjects, a system had to be devised to supplement the printed inventory lists of the library's contents. Arranging materials by subjects seemed to be the obvious conclusion and in 1605 Francis Bacon divided all human knowledge into three kinds of science: history (memory), poesy (imagination), and philosophy (reason) (Lerner, 2009, p. 120).

These three major categories were added to and subdivided over the years as new subjects not envisioned by the original author were being written about. The other main challenge to this evolving classification scheme was that libraries, each one operating independently, were inconsistent in how they applied the categories. There remained the need for a standard system of classification that could be readily adapted by many libraries to promote uniformity and efficiency.

One of the great pioneers in this field was Melvil Dewey (1851–1931), the father of the Dewey Decimal System, used by many libraries throughout the world. Dewey developed a system in which each subject classification was broken down by a

Selecting and Implementing an Integrated Library System.

numerical code with further subdivisions under each class. Under this system, each book was assigned a unique call number, making it easy to shelve and retrieve, all grouped within the same subject area. This was a huge improvement over previous systems and allowed libraries to more easily accommodate the growth of their collections and the introduction of new subject areas. This idea of dividing the world of knowledge into increasingly complex subject areas led to the more specialized Universal Decimal Classification and the Library of Congress (LC) Classification System.

It is at this point that the first card catalogs were introduced, with each book having an individual card in the catalog. Now, instead of consulting a dated printed list of library titles, patrons could look up their favorite materials in a catalog that was not static but could be added to indefinitely as new materials were purchased for the library. With the wide publication of the Dewey system and the standardization of cataloging, other libraries too began to adopt the same classifications for their materials.

Since the LC was serving as the unofficial national library of the United States, it had one of the largest collections and staffs to catalog materials for its collections. In 1902, the LC began selling copies of its printed catalog cards to other libraries, saving individual libraries the expense of having to catalog materials already owned by LC (Lerner, 2009, p. 179). This was one of the first steps in terms of library automation, even though it involved printed materials and was well before modern technology entered the picture. The idea of library cooperation and resource sharing were slowly becoming one of the cornerstones of how libraries operated (Figure 1.1).

The development of computers led to the next steps in library automation, mirroring what was happening in the rest of the business and government information world. Society was being changed by the advent of computing and libraries wanted to take advantage of this new technology to help them manage their growing collections, especially as the economy and funding for libraries grew during the 1960s.

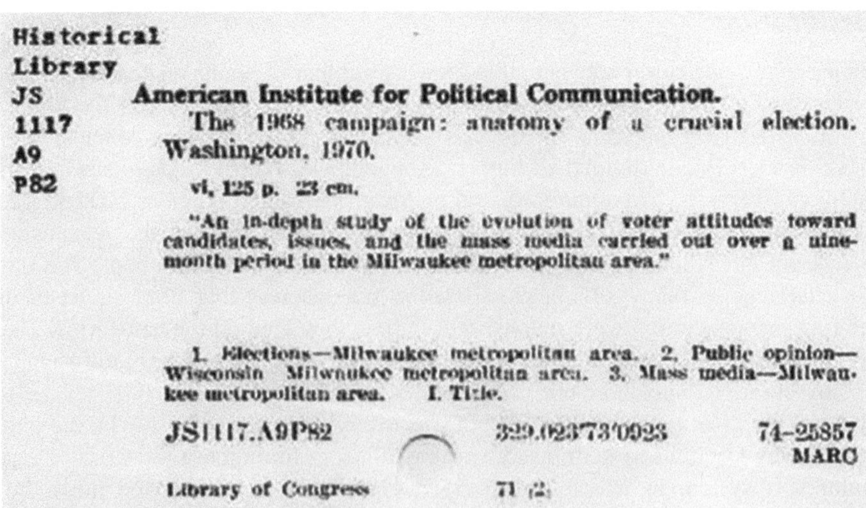
Historical
Library
JS American Institute for Political Communication.
1117 The 1968 campaign: anatomy of a crucial election.
A9 Washington, 1970.
P82 vi, 125 p. 23 cm.

"An in-depth study of the evolution of voter attitudes toward candidates, issues, and the mass media carried out over a nine-month period in the Milwaukee metropolitan area."

1. Elections—Milwaukee metropolitan area. 2. Public opinion—Wisconsin—Milwaukee metropolitan area. 3. Mass media—Milwaukee metropolitan area. I. Title.

JS1117.A9P82 329.023'73'0923 74-25857
MARC

Library of Congress 71 (2)

Figure 1.1 Example of LC card catalog card.

As library collections grew, the necessity of finding a way to be more efficient became even more acute, especially with the labor-intensive aspects of library operations. One of the most expensive sources of labor costs for most libraries was the cataloging operation, in which each item had to be described, both physically and intellectually, before it could be added to the collection. Creating an original cataloging record required highly trained staff and time-intensive procedures. Multiply this cost by the number of libraries with their own team of catalogers and it is readily apparent that this cost would be difficult to maintain.

As has been already noted, the development of shared classification schemes allowed libraries to share cataloging records with other libraries through printed cards. But with the advent of automated library systems, it became possible for libraries to actually share their catalog records directly with each other without waiting for a third party supplier to produce cards and mail them out. The development of automated systems for the sharing of bibliographic records resulted in the development of bibliographic utilities, an important milestone in the library automation world.

The most famous bibliographic utility is OCLC, established in 1967. Originally conceived as the Ohio College Library Center, a statewide bibliographic database, it quickly grew beyond its geographic area to encompass bibliographic records for other national (and later international) libraries. It was followed by the Research Libraries Information Network (RLIN) and the Western Library Network (WLN) both of who were serving a similar function to OCLC.

The establishment of these bibliographic utilities would not have been possible without the development of bibliographic record standards, as each library would need to adhere to a common standard for sharing records. Pioneered by the LC in 1961, the MAchine Readable Cataloging (MARC) format was developed and quickly became the standard for most American libraries. The adoption of MARC standards meant that libraries could contribute to and copy records from the bibliographic utilities to their local catalogs, secure in the knowledge that all records met the agreed-upon standards. This was one of the most important developments in library automation as one of the most expensive jobs in the library operations was greatly simplified and expedited through the sharing of bibliographic records on a global scale (Figure 1.2).

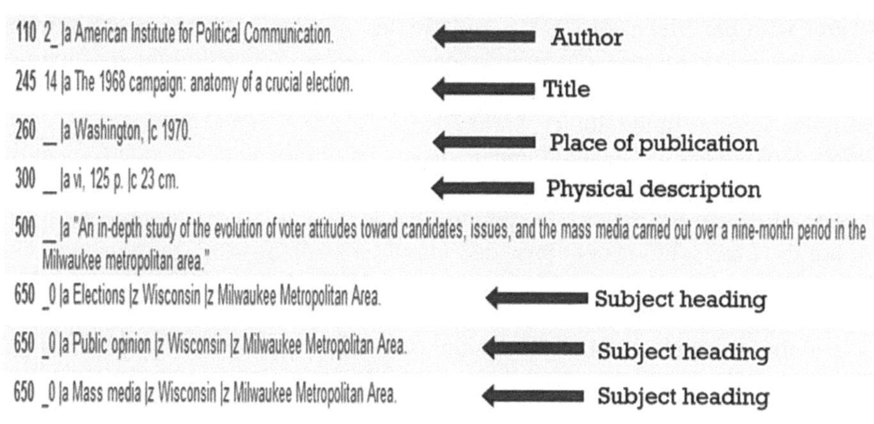

Figure 1.2 Example of MARC record.

Following the automation of the cataloging operation and the continued growth of the computer industry, other aspects of library operations became targets for automation. Although some libraries had experimented with punch card systems for circulation purposes, most library record-keeping was still done manually, with a mixture of professional and nonprofessional staff. Starting with the technical services operations, the Northwestern Online Total Integrated System (NOTIS) was developed by Northwestern University in 1968 as the first integrated library system. This system was one of the first to link circulation functions and technical services functions in one unified system, followed by the development of an online public access catalog in 1985 (Figure 1.3).

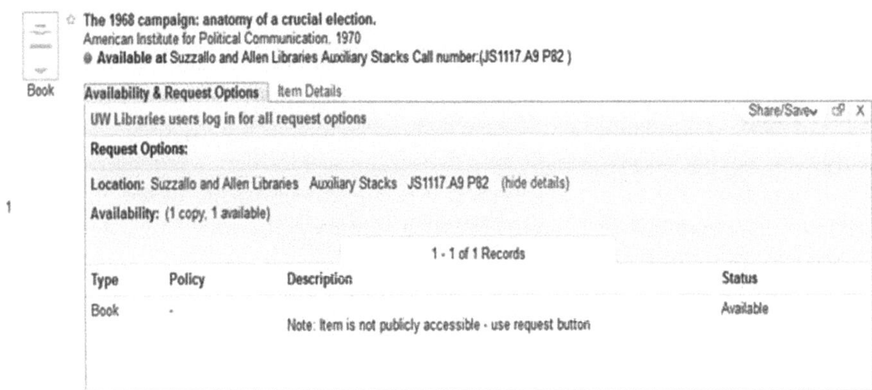

Figure 1.3 Example of online catalog record.

The success of NOTIS and its availability as a library system for purchase launched a revolution in the library world. No longer did each library have to have an in-house programming staff and information technology (IT) support to develop its own library software; with the advent of NOTIS and similar systems, a library could purchase software for its operations from an independent vendor. This led to the development of an automation marketplace of many vendors, offering a variety of integrated and stand-alone systems to the library world.

When looking back at the more recent history of library automation, it is clear that for over 100 years, librarians have been attempting to discover ways to use their current state-of-the-art technology to expand library services. Often these technologies were not the final solutions but provided instead a stepping-stone to the adoption of solutions that took advantage of new technological developments. Some of the highlights from library automation history are shown in Figure 1.4.

The ability to automate library functions with an integrated library system is only possible when the library has data stored in an electronic format that can be utilized by the system. Many libraries made the transition from a paper-based storage system

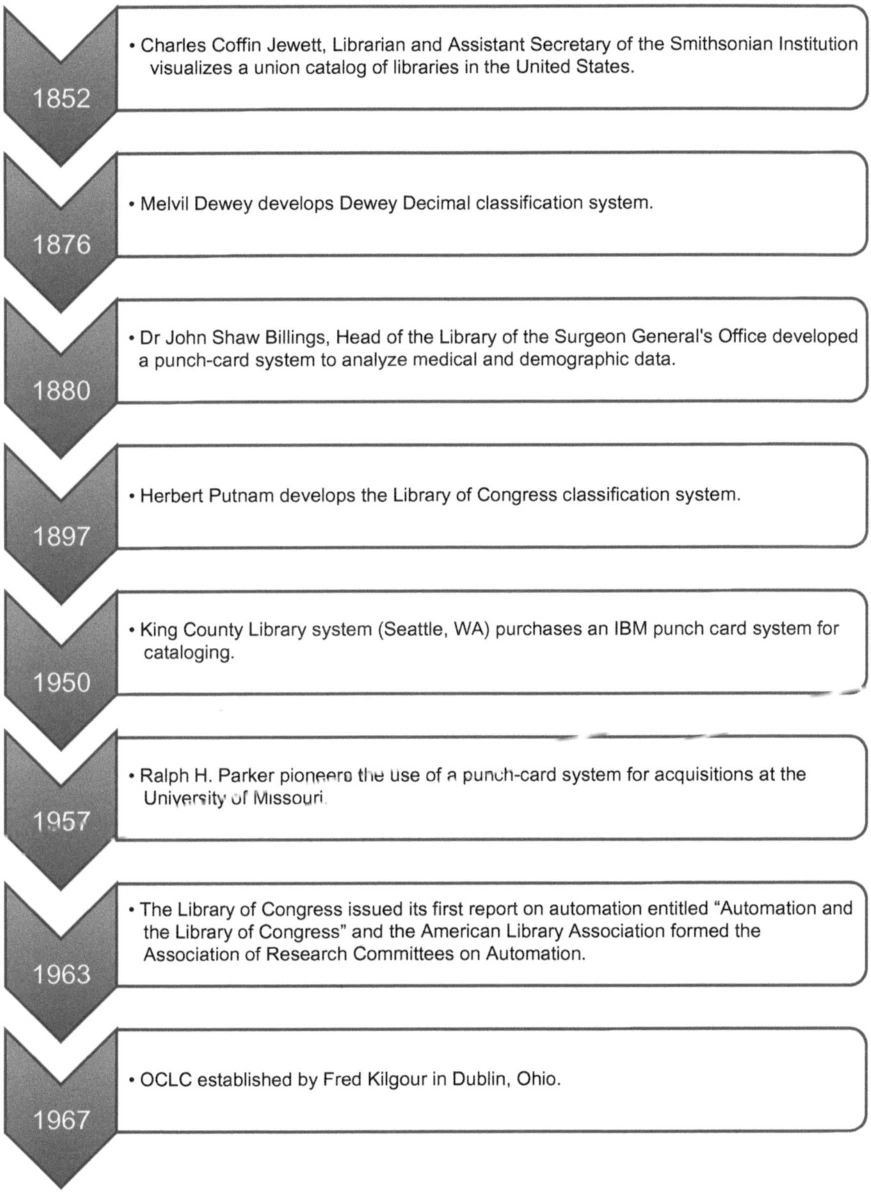

Figure 1.4 Milestones in library history.

to electronic records upon the purchase of a new system, necessitating new workflows and procedures. A complete understanding of the record types used in modern libraries and examples of typical workflows is critical to anyone wanting to be familiar with the development of integrated systems.

Record Types and Print Library Workflows

2

Components of library operations

Before beginning a discussion of library automation, it is necessary to first understand the various library operations that are involved. Although each library will be different and may have a unique organizational structure, the basic outlines of standard library services can usually be found in every organization. They are traditionally split between technical services and public services (often now called collection resources/management and access services), but have many of the same characteristics across the library spectrum.

Technical services

In this area, the traditional subunits are cataloging (often called metadata), acquisitions, serials, binding, and processing. Most units in technical services (cataloging, acquisitions, and serials) usually deal with print, electronic, and other types of formats that the library collects, while the binding and processing is mainly devoted to print resources. In some libraries, the electronic resources may be split into their own unit (often called database support).

Public services

This is the area that includes subunits such as circulation, interlibrary loan (often called resource sharing), and the online catalog or discovery layer. Circulation is a function that is often tied into the technical services workflow, as it is primarily dealing with the print resources. The resource sharing function also deals with the print resources for the most part and is a natural partner with the circulation functions.

Library records

The fundamental mission of libraries is to preserve and make available library collections in multiple formats. To accomplish this task, libraries have developed a set of records to help them with their task, which encompass the types of information that a patron would need to retrieve the requested material.

Selecting and Implementing an Integrated Library System.

The first, and for most libraries the most important, type of record is the bibliographic record. This record, usually created by professional catalogers (although increasingly becoming available from publishers and vendors), represents the intellectual building block of the library. The bibliographic record is the attempt to describe the resource (whether it be a book, DVD, database, or eBook) completely and uniquely. It contains all the information about the resource broken down into standard fields based on the MARC format.

In most library systems, the bibliographic record is the main record and forms the basis of the catalog. It is the record that most patrons are familiar with as the search terms that they input into an online search engine (or formerly used with a print card catalog) come directly from the bibliographic record. Title, author, subjects, etc., are all fields from the bibliographic record which are accessed by the various indexes that the library has created for their system. When most patrons think of a library catalog, it is the bibliographic record that is most often associated with that mental picture (Figure 2.1).

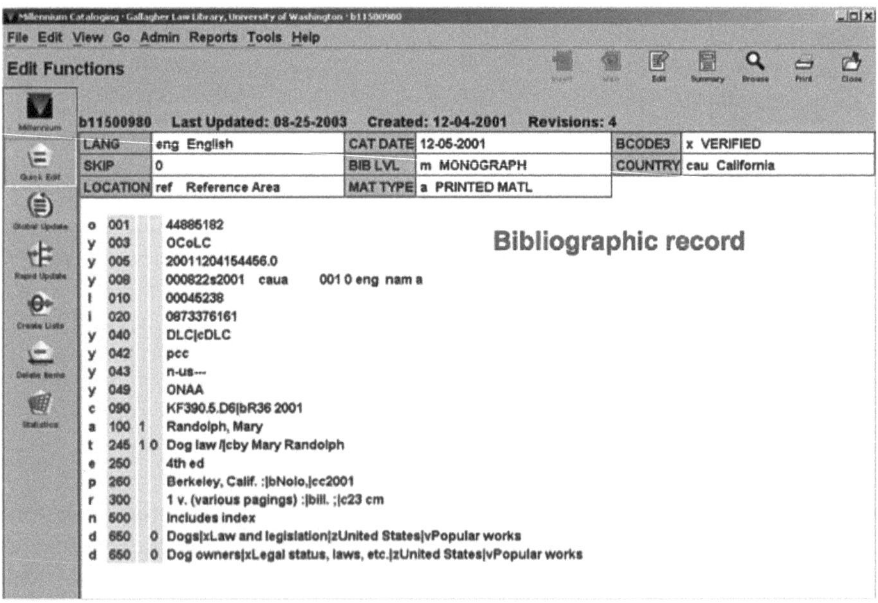

Figure 2.1 Example of bibliographic record.

There are, however, other types of information that the library needs to track which are not necessarily appropriate for the bibliographic record format.

The order record is one of the first records created because it begins the workflow process of obtaining new materials for the library. Generally, the order record is attached or linked to the bibliographic record in some way to use the bibliographic fields (author, title, publisher, etc.) as part of the order record. In addition, the order record is intended to transmit specific information to the vendor or publisher from whom the library is purchasing materials.

The order record should contain all the information that is necessary for the vendor or publisher to identify the resource being ordered as well as any instructions on how to complete the order. It should also include necessary information for the staff to use to process the order. So, for example, the order record will have fields that detail the vendor to whom the order is being sent, the selector who chose the resource, the budget fund that is being used for purchase, the number of copies desired, as well as any special instructions (e.g., hardcover versus paperback, type of binding, signed copy) (Figure 2.2).

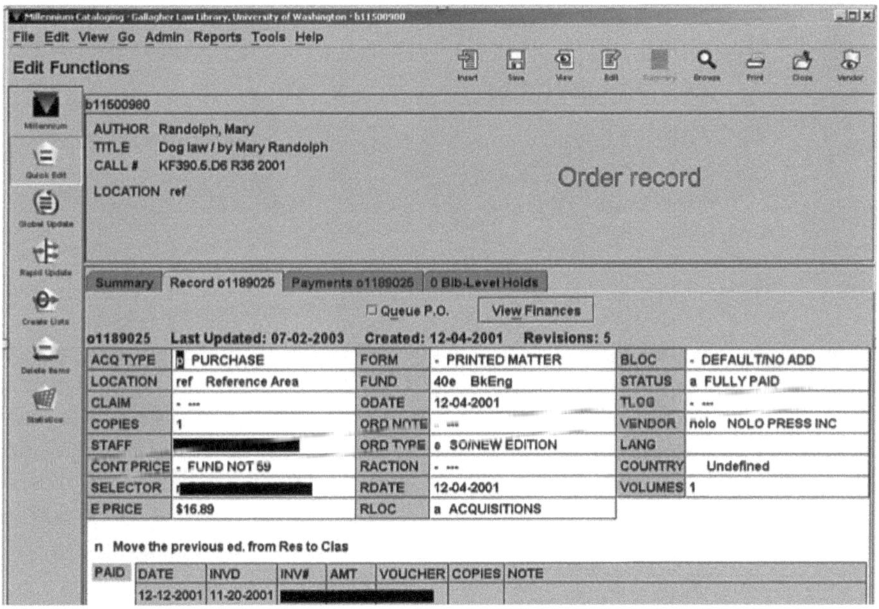

Figure 2.2 Example of an order record.

The item record is also a record type closely associated with the bibliographic record but with a different purpose. While the bibliographic record represents the *title* that a library owns, the item record represents the individual *copies* that the library owns. Thus, a library can have many copies (or volumes) associated with a single title. The title may be unique but each copy is singular.

Most item records are used as the basis for the circulation of materials. In addition to the individual copy or volume number in the record, the item record also usually has the item barcode (unique to that resource), the loan rule under which the resource may or may not circulate, and the patron type (which patron category may borrow the resource). The item records are linked to the bibliographic record (think of a parent–child relationship with the bibliographic record as the parent and the item(s) as the children). Together they provide the patron with all the information that is needed to both search for a resource and determine if it is available for borrowing (Figure 2.3).

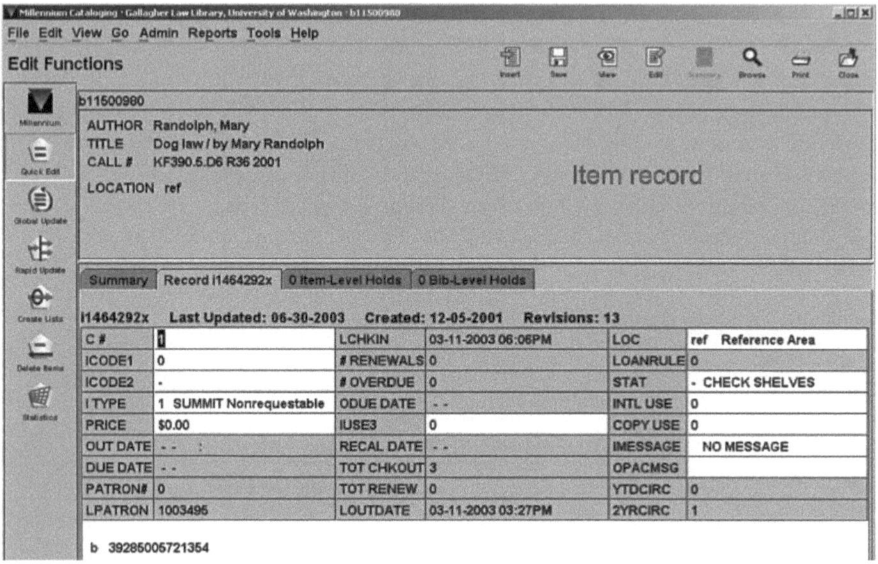

Figure 2.3 Example of an item record.

However, before this resource can actually be borrowed from the library, another type of record is necessary—the patron record. Each library patron who is eligible for borrowing privileges has to have a patron record in the automated system. The patron record lists the contact information for the patron (name, address, phone number, email) as well as the type of patron (student, faculty, public, etc.) and an individual identifier for that patron. That identifier can be a barcode, a numerical identifier, email address, or some other form of identification that can be used in the system (Figure 2.4).

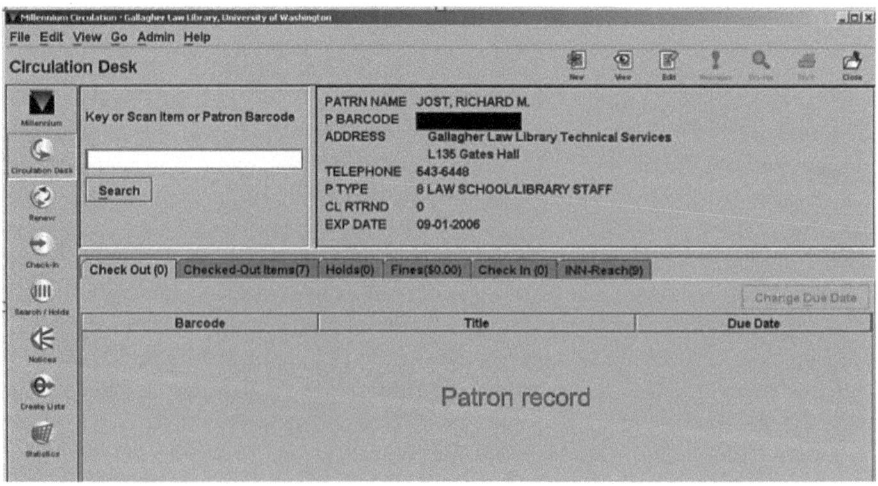

Figure 2.4 Example of a patron record.

For some resources, especially serials, there are other types of records that are relevant. Like a simple monograph, the serial resource needs a bibliographic record to describe it and item records to circulate the various issues, volumes, or pieces that are published under its title. What it also needs, however, is a way to track these individual pieces as they are received in the library, which, in most cases, is done through the use of serials (commonly called checkin) records. The checkin record will allow the library to continually receive the serial resource as it is published, recording each piece in one place. This makes it easier to track what has been received in the library and more importantly, to be aware of any pieces that never arrived and need to be claimed from the publisher or vendor.

The checkin record is also usually the place where the information about a library's entire run of a title is held. This information, often called a "holdings statement," is the description of when the library subscription began (volume, title, year) and the current status of the subscription (last issue received if still an active subscription or, if canceled, the date and number of the last issue received). This holdings information is usually displayed in the public catalog for the patrons to determine if the library resource they are looking for is current or whether it has been canceled and no longer received. The holdings statement can also list any gaps in the pieces received or published and is especially useful for other libraries processing interlibrary loan requests. Quite often systems will have a separate holdings records where this information can be stored, usually in MARC format (Figure 2.5).

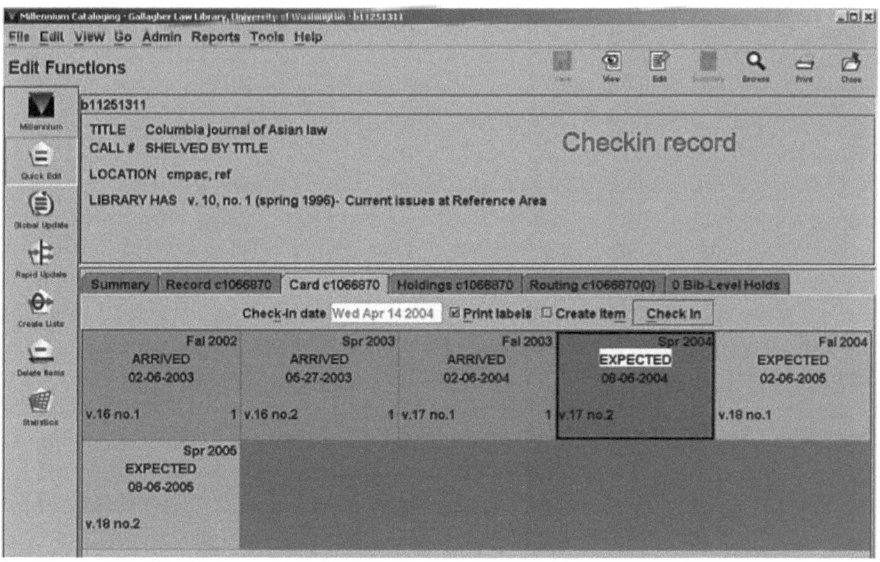

Figure 2.5 Example of a checkin record.

Library workflows

To the uninitiated, it may appear that library materials "magically" appear on the shelves, requiring little library staff intervention. Those who work in libraries, however, know that there is a complex set of workflows for every library transaction. These workflows are often the reason that many libraries want to automate in the first place, as they can be very labor-intensive and cumbersome if done strictly through manual procedures. Because these workflows will form the basis of many integrated library system projects, it is important that we understand the various steps.

Technical services

The workflows in technical services are areas that lend themselves to automation very readily. Most, if not all, integrated systems have automated these procedures; indeed, many system vendors got their start with projects involving the technical services area.

The first step in the workflow is collection development, the process whereby the library decides what to purchase for the collection. This decision can be made by a variety of staff. In some cases, the director makes all the decisions while in others designated staff (usually called selectors) is given responsibility for selecting in certain subject areas. A library can also use a committee to make decisions, designated members of the reference staff, or select any combination that best meets their needs.

The purpose of this selection decision is to acquire materials to best meet the needs of the library's patrons and that fulfill its mission statement. The actual process can be done in many ways—suggestions from the patron community, sales information from vendors and publishers, online reviews, exhibit hall sales information from professional meetings, etc. It is the job of the selector to sort through all the possible sources of new materials and decide which ones would be best for his or her individual library collections.

Although there may not be as much automation in this first step as in some of the others, there are tools available for the selector to use. Many vendor sites have online reviews and bibliographic descriptions that provide more information for a selector to determine if a resource is appropriate for his or her collection. In addition, many selectors may use OCLC (or another bibliographic utility) to gather more information through a full cataloging record for the resource as well as to see what other libraries already own the resource. In addition, many libraries belong to a vendor approval plan, where relevant materials are sent automatically to a library based on a collection profile. Any of these methods and many more go into the work of selecting materials for building a library collection.

The second step in the workflow is the acquisitions process, which is the step that translates the decisions by the selectors into an actual order for purchase. Many people, even in the library world, use the terms "collection development" and "acquisitions" interchangeably, but they actually represent two different processes. Collection development is the intellectual process of deciding how to build a collection, while acquisitions is the more mechanical procedure of procurement and recordkeeping. In many libraries, the same person may do the two processes, but they do represent two distinct endeavors.

The acquisitions process is where one traditionally finds the beginning of an integrated library system workflow. In this process, the order record is created, often the first record for a resource that is to be added to the collection. In most cases, the order record has to be linked to a bibliographic record, as the standard bibliographic fields (author, title, edition, subjects, etc.) are usually not fields in the order record. Many libraries will download a bibliographic record from OCLC to serve as the host record for the order record while others may just create a brief bibliographic record on their own to be updated at a later stage.

Libraries that use an integrated system will most likely have an order record template that they use to prepare orders. This template will include all the fields that the library needs for its orders and can often have repeatable fields (e.g., the address of the library) that are always included when beginning to input information into a blank template. Setting up these fields ahead of time can save many hours of typing and ensure greater accuracy.

Once the order record is complete, it is sent to the vendor or publisher for order fulfillment. Most libraries use email to send their orders, but they can usually also be printed out and mailed if they do not have electronic access. Another way of ordering is to order resources directly on the vendor or publisher web site, often by making selections from a title list or by using a vendor like Amazon. In these cases, the library must still create an order record in the integrated system for these items, even if an actual order request is not sent to the vendor. There still has to be a way to track these orders and to maintain records for them upon arrival in the library's system. Libraries may also subscribe to approval plans from vendors, where subject profiles are used to send materials "on approval" to the library for possible purchase for the collection.

Most integrated library systems will supply a unique order record number for each order as well as a purchase order number, often used in the payment process. The order records can usually be displayed in the public catalog, if desired by the library, to let the patrons know that a resource has been ordered but not yet received. Often, these records are brief displays that provide some bibliographic information about the resource and the date that the resource was ordered. Some libraries prefer to not display this information to the patron community but keep it on the staff side of the integrated system until it has been received and is available for patron use.

Once the resource has been received in the library, the acquisitions staff is usually the staff that opens the boxes, checks to make sure that that resource that has been sent is correct and undamaged. If there is a problem with the order (e.g., wrong number of copies, incorrect edition, damage to binding), the acquisitions staff will contact the vendor or publisher to report the problem(s) and seek a solution. Once the acquisitions staff is satisfied that the order has been filled according to specifications, they will pass it on to the next step in the process.

The third step in the workflow is cataloging. The catalogers are the staff who will create the cataloging record for the resource, enabling it to be displayed and searchable in the public online catalog. These records, usually in the MARC format, form the basis of the library database and represent the scope of the collection.

Once they receive the resource from the acquisitions staff, the catalogers will need to decide if the bibliographic record that was used in the ordering process (the host bibliographic record) is sufficient for the online catalog. As has already been seen, many of the bibliographic records used in ordering are brief records, created just for the ordering process: they lack key fields, like call numbers and subject headings, which are necessary for searching and retrieval. Using OCLC (or another bibliographic utility), catalogers will search the utility for a full cataloging record, that is, a record with the most complete information. Depending on the integrated library system, it is usually an easy process to download the complete record from the bibliographic utility to the local database and overlay the brief host bibliographic record. Catalogers may make changes or additions to the record that they download or they may accept it as is with no changes.

The process described above is called "copy cataloging"—a record is copied from the bibliographic utility into the local database for local use. But what happens when there is no record in the bibliographic utility for the resource that has been ordered? In that case, the catalogers must perform "original cataloging" by which they create a full bibliographic record based on the resource that they have in front of them. As you can imagine, this involves much more time and effort than copy cataloging, resulting in higher labor costs and a slower rate of production. Even though a library has to pay a membership fee to join a bibliographic utility, the fee is usually a cost savings when compared to doing original cataloging, especially for large and complex collections.

Once the catalogers are finished with the resource, the material proceeds to the fourth step in the workflow, which is usually called processing. This is where the third record type is created for the resource, the item record. The item record will contain the copy number, the item barcode, the loan rule under which it circulates, special notes (e.g., "check for CD in book pocket"), and any other relevant information. These item records (each copy must have its own item record) are linked to the bibliographic record and exist in conjunction with the order record that started this process. It is also at this stage that the resources are marked with the library identification through stamps or labels, call number labels are affixed to the spine and security tags are applied. Once all these details are taken care of, the resource is ready to be sent to the circulation department for shelving and patron use.

There is another record that most integrated library systems have that is also a vital part of library record-keeping, the serials (or checkin) record. Suppose that instead of a print monograph, the library was ordering the first issue of a brand-new journal, along with a continuing subscription. The same first four steps of the workflow would need to be followed: (i) collection development, (ii) acquisitions, (iii) cataloging, and (iv) processing just as if it were ordering a print monograph. The extra step for any serial publication, however, is not what to do with the first issue received, but what to do when the second issue arrives? The library has to have some way of tracking the individual issues or volumes that are published and received in the library and the serials record is the record most often used. For this purpose, the serials record, based on the old kardex paper record system, is used to record the receipt of individual pieces of the subscription as they arrive by date and/or payment. The serials record is also extremely useful for claiming materials that are not received but should have been, as it alerts the staff member when an expected shipment is late or has not been received. This serial record joins its siblings (the order and item record) and is linked to the bibliographic record for the resource.

This is a very labor-intensive process, even with the aid of an integrated library system for record-keeping. The complexities of managing this multistep process in a manual system would be daunting. Although these steps may be called something different in each library or may be altered in the order in which they are performed, the general outline of this process will be found in most library operations (Figure 2.6).

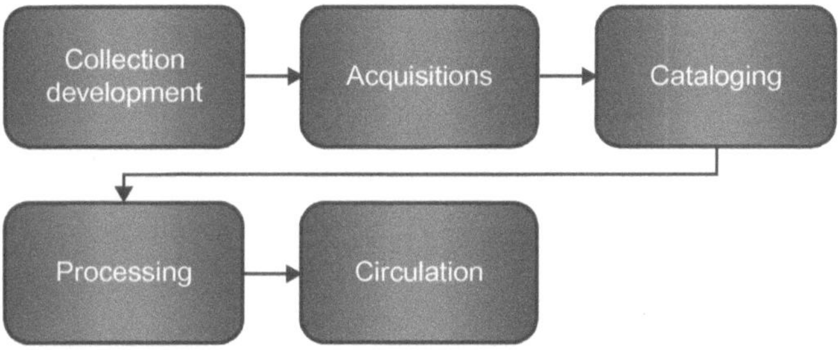

Figure 2.6 Print workflow summary.

Public services

Once library materials have been fully processed by the technical services staff, they are ready for patron use and circulation (or access if they are electronic resources). There is no need to create a separate "circulation record" as the item record (already created as part of the technical services processing) is the record with the relevant information for circulation.

Once a patron (with a current patron record) brings a resource to the circulation desk to be checked out, the intersection of the item record (can these materials be borrowed?) and the patron record (is this patron in good standing and allowed to borrow this resource?) come together to complete the circulation transaction.

Discovery

The function that is really a separate operation and one of the most critical in an automated system is discovery. In many systems, the discovery function can be decoupled from the technical services operations. It is very common to purchase the two parts of an integrated library system from two different (and often competing) vendors.

At this stage, it is important to differentiate between the traditional online public access catalog (OPAC) and a discovery platform. An OPAC in most libraries contains the electronic records of materials that the library actually owns and that are housed in the library. This is the same service that the traditional library catalog provided in the non-electronic age, although the advantage of the online catalog was greater search capabilities and links to subject headings, authors, and publishers. Still, this was, for the most part, a closed universe: the catalog would search for materials within the library but no further.

As the library world became rich in a variety of electronic resources, library catalogs became just one of numerous resources available to the library patron. The challenge at this point in time, though, was that patrons had to search each resource (or silo of resources) individually. Patrons had to determine for themselves whether the information that they were seeking was contained in the OPAC, in a research database, in a digital collection, or in an electronic journal. In addition to knowing which resources to search, a further complication was learning the various interfaces for each resource. Based on the number of resources being searched, this task could be quite challenging, even for experienced library users (Figure 2.7).

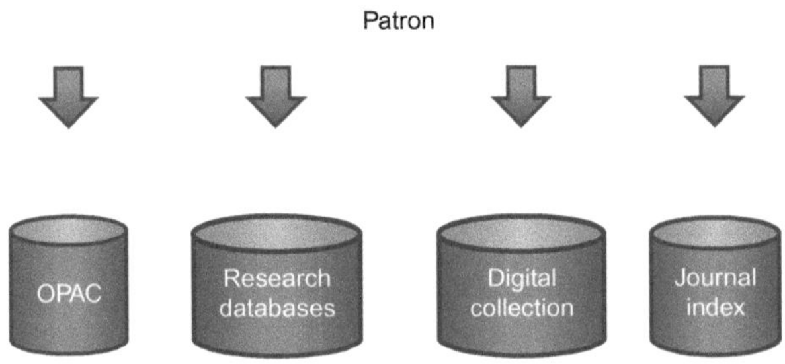

Figure 2.7 Database searching pre-discovery platform.

A discovery platform contains all the same functionality of the traditional online catalog, but also searches a much wider universe. Depending on what online resources a library subscribes to, a discovery platform can search databases and other online resources along with the library's collection in one search. Instead of the patron having to know which database might have materials that would be helpful to him or her and mastering each particular database's interface, the discovery platform allows one search and one set of interface instructions to retrieve a much more varied list of results. These results, presented to the patron in one browse list, come with access information for the online resources or the physical location within the library. If the library happens to be a member of a consortium, often these materials also display in the search results list, along with links to initiate a borrowing request from another institution (Figure 2.8).

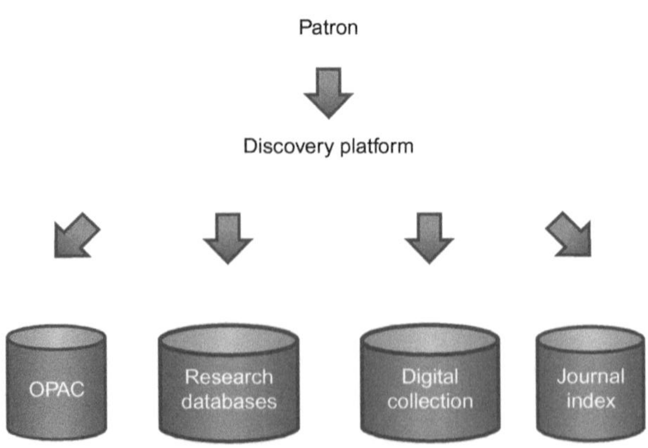

Figure 2.8 Database searching post-discovery platform.

Many vendors sell discovery platforms to use either with their own integrated library system or with the system from another vendor. These discovery platforms are usually quite customizable and can work with a variety of online resources and databases. When a library purchases a discovery platform, the online catalog becomes just one of the databases that is searched by patrons along with the other outside resources, which the library does not own but has access to as part of its online subscription collection.

A discovery platform has enormous potential to help bring relevant resources to patrons with one search. After retrieving the results from a search, the patron can often refine the results list through the use of facets, limiting by date, author, full-text, publisher, etc., or whatever other facets that the library has set up for its use. For those wanting to restrict the search to just the physical items in the library, there is often a facet that will allow patrons to screen out online resource and focus on just certain types of formats (e.g., print resources) as a way to locate materials.

The discovery platform is the piece that unites both public and technical services. Instead of the rigid separation of these departments in traditional print-based libraries, technology has knit them together in a new service model, where all library staff members now are part of the wider definition of public services operations of a library. Just as technical services staff members have to understand how the records they create are being used by patrons to access materials, so too do public services librarians have to understand how the records are constructed if they want to help patrons navigate the library collection. The breaking down of barriers in the library organization through the use of technology is one of the most important developments in the world of library automation.

The records and workflow for print resources have been a part of library operations since library collections began. With the advent of electronic records for print resources, procedures may have changed but the basic concepts of the workflows remained approximately the same. The real change, however, came with the advent of electronic resources, which had no print counterparts. Those materials which were "born digital" presented a new challenge to libraries and the integrated library systems they used for other resources.

Electronic Resources

3

With the increasing emphasis on electronic resources, the need for an integrated system that can handle this format is paramount. The workflow for electronic resources is different from that of print resources in several important ways and is not often accommodated very well in a system built strictly for print. To meet this need, most library automation vendors offer either systems that are specifically geared to electronic resources or have a component in their existing system dedicated to these resources.

The most common of these components is usually designated as an electronic resource management (ERM) system that stores and tracks information about electronic journals, database, and electronic books. Although it is possible to track this information in print-based records, it is often unsatisfactory as the information in electronic resources does not match the fields in a print resource record. Libraries either have to add lengthy staff notes to print resource records or rely on an outside source (such as a spreadsheet) to fully track these resources. The better solution is to invest in an ERM which is specifically tailored to these formats.

The record in an ERM (often labeled an "erecord") contains fields that are exclusive to electronic resources, such as access information, authentication methods, and the resource uniform resource locator (URL). Often, these fields are displayed in the catalog for the public to see while more information such as the subscription price, Internet protocol (IP) range, and subscription license are only visible to the staff. Below is a sample of an erecord from the Yale Law Library catalog (Figure 3.1).

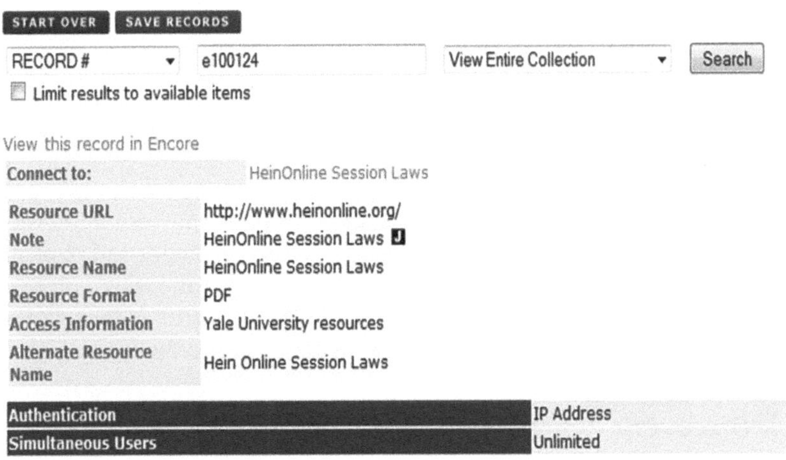

Figure 3.1 Example of an ERM record.

Selecting and Implementing an Integrated Library System.

The ERM has many benefits to the library as it:

- tracks electronic resources from point of order through licensing and final access;
- provides information about the data providers, consortial arrangements, access platform;
- provides contact information for all content providers;
- logs problems with resources and providers;
- links license documents to resource record.

Electronic resources life cycle

In addition to an ERM, the library often has a separate workflow for electronic resources which differs from print resources in several ways. Some of the steps are the same (e.g., creating an acquisitions order record) while others (e.g., setting up sample trials for evaluation) are only relevant for electronic resources.

The steps for the electronic resources workflow are outlined in the following sections.

Step 1: Evaluation

- Use subject specialists for recommendations
- Set up sample trials for evaluation
- Compare to other databases in library collection for overlap
- Solicit recommendations from staff and patrons

Collection development is still a necessary step for electronic resources. The library still has to go through the intellectual process of deciding which electronic resources are appropriate for their collection, just as they would for a print resource. The same considerations on relevancy, patron demand, cost, and support of the mission statement must be taken into account.

Step 2: Acquisition

- Create order records in library system for order management
- Create ERM record for license management
- Pay invoice

The acquisitions step is still part of the process for electronic resources. No matter what type of resource one is buying for the collection, the vendor and publisher still need an order record to activate the resource and the library still needs a place to track its orders and expenditures. This is where the ERM record is created to record some of the information on the electronic resource for the staff as the order is being processed.

Step 3: Discovery

- Create bibliographic records for discovery
- Set up proxy for outside access

- Set up database portal on web site
- Set up training schedule for public service staff
- Publicize to patrons and staff

The cataloging of the electronic resource is often still part of the electronic resource procurement process. In many cases, the library will add a bibliographic record for the electronic resource to the online catalog. Some libraries just rely on separate listings of databases to which they subscribe posted on their web sites, but most have a record that can be used in the catalog for searching to alert the patron to the existence of this online resource.

Step 4: Manage

- Troubleshoot problems (broken links, access problems)
- Help patrons use the resource
- Monitor resource status and availability

Step 5: Review

- Compare usage with statistical information
- Determine cost per use
- Compare overlap and other subscription options

It should be remembered that the higher cost of most electronic resources can be partially offset by the savings in technical service processing necessary to prepare them for patron use. Since there are no individual copies to check out (no need for item records) and no individual volumes being received (no need for serial records), staff time can be saved. There will also be savings in the circulation area as electronic resources do not need to be checked out at by circulation staff.

Step 6: Renew/cancel

- Communicate with vendor on decision:
  - Renewals:
    - Renew licenses
    - Update IP ranges (if necessary)
    - Pay invoice
  - Cancellations:
    - Notify colleagues and patrons
    - Remove from library web site and catalog
    - Publicize replacement service (if available) (Figure 3.2)

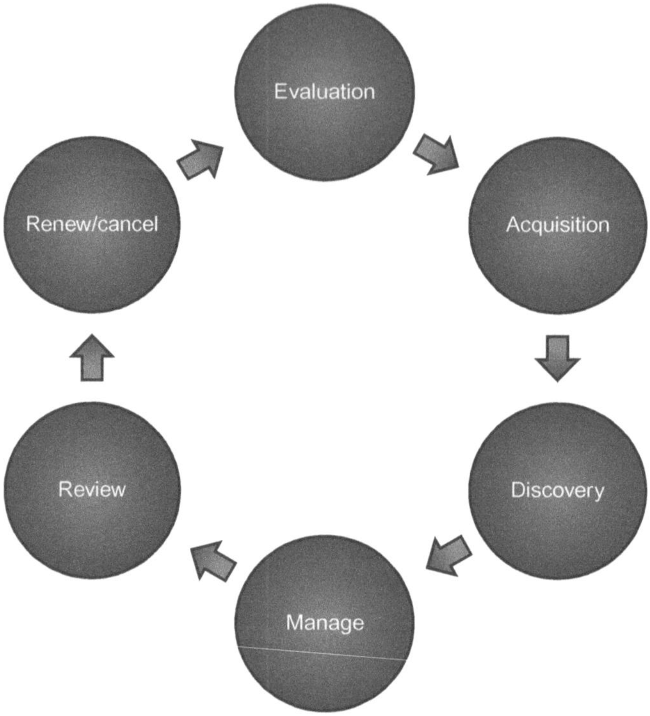

Figure 3.2 Electronic resources life cycle.

One tool to use in reviewing electronic resources (Step 5) is Counting Online Usage of Networked Electronic Resources (COUNTER), an international initiative that sets standards that facilitate the recording and reporting of online usage statistics in a consistent, credible, and compatible way. The benefit of using COUNTER is that it can compare usage statistics from different vendors, allowing the library to develop metrics to determine the cost per use of individual database services. Depending on the outcome of the COUNTER analysis, the library has an important source of information to make more well-informed library purchase decisions.

For many libraries, using demand-driven acquisitions (DDA) is another method for procuring electronic resources, in this case eBooks, for the collection. Instead of library staff selecting the individual eBook titles to add to their collection, the libraries subscribe to an eBook package from a database vendor. The DDA program allows libraries to make eBooks discoverable to their patrons (via the discovery platform) prior to actual purchase. Once a title has been accessed a set number of times (determined by the contract with the vendor), the eBook will be triggered for purchase by the library. This places the selection decision on patron use and popularity, ensuring that the eBooks that are added to the collection meet patron needs.

One of the challenges with electronic resources is providing patron access. Unlike print resources which are either on the shelf or not, the library must have a robust electronic infrastructure that allows patrons easy and quick access to their electronic resources at all times. Problems with access to these important resources may result in a forfeiture of use at a considerable cost to the library budget.

To alleviate this problem, most libraries use some sort of link resolver to facilitate access. After a patron has done a search for an article citation or journal in the library's discovery platform, the link resolver compares the search results to the library's electronic databases and displays links to the electronic resources that contain the article or journal. These links (often for full-text) may be customized by each library depending on what electronic materials the library has in its collection. The patron will often get multiple links to the same content as different vendors may contain the same resources in a variety of databases.

The complexities of print and electronic workflows and record-keeping require a sophisticated knowledge of the role that integrated library systems have in the modern library organization. Knowing which system will be the most efficient and effective for an individual library can be a daunting task for those without specialized training in the systems area. Finding and hiring experienced systems librarian may be one of the most important factors in a successful library migration to a new system.

Systems Librarians

4

Systems librarianship

One of the most important aspects of any systems projects is finding the person who will be the main architect of the project. In most libraries of any size, the system project will be a collaborative process, but there is usually the need for one person to be the coordinator or manager of this type of project.

Often, the responsibility for leading a system project falls onto the shoulders of someone who is already employed at the library (e.g., the head of technical services, the senior cataloger, or the head of circulation) who may or may not have the necessary qualities to succeed in this type of project management. In addition, current staff members also usually maintain their ongoing set of job duties, making the burden of trying to manage a new systems project very challenging.

To mitigate this burden on existing staff, it is advisable (if possible) for the library to hire a dedicated person to manage an automation project. Although there are many titles for this set of job duties, the most common one is "systems librarian," although these job duties may also be included in the job description of an "emerging technologies librarian" or "electronic resources librarian." Often the latter two titles refer to someone dealing more with databases, eBooks, and other electronic resources, but the field is still pretty fluid in assigning titles and job responsibilities.

What benefits would the library get from hiring a systems librarian for its automation project? To begin, the library would have a staff member specifically dedicated to this project. The systems librarian's major responsibility would be to shepherd the automation project to a successful conclusion without the distraction of being pulled away by competing job duties. This would allow the systems librarian to be focused on the project in a way that will have the greatest chance of success, as he or she will be available to handle the multiple seen and unforeseen issues that may occur.

The other major advantage to the library in hiring a new systems librarian is the ability to hire someone with the proper background and credentials for leading an automation project. As stated above, making an existing staff member in charge of the project may seem like a logical move as he or she will have a long history with the library and the library collection. This will definitely be an asset but there are other even more important skills that a systems project manager must have to be successful. Hiring the right person for this job is one of the most critical decisions that the library can make at this early stage of the project.

Definition

Before proceeding, it would be helpful to have a definition of this new staff position. A very good definition can be found in Wilson (1998): "Systems Librarians specifically concentrate on supporting library operations through the application of technologies... They seek to manage the organizational technological resources in a manner that best serves all departments within a library" (p. 11).

Depending on the size and complexity of the institution, this could encompass a wide range of activities. In a small library, the systems librarian, in addition to being responsible for the library automation project, may also have job responsibilities in technical services, circulation, or reference. In a large organization, the systems librarian may be part of a library systems department, responsible for library automation but also for any other software and hardware used throughout the library. It is almost impossible to come up with a "typical" systems librarian job description as it can vary so widely between different types of libraries but there are commonalities in the skills that a systems librarian should have that will link them no matter where they are employed.

Skills

To be successful as a systems librarian, there are definite qualities that a library should seek in any candidate for this job position. Wilson (1998) lists qualities that he suggests that systems librarian should have:

1. Flexibility and balance
2. Sound judgment
3. Curiosity and risk taking
4. Follow-through and persistence
5. Time management
6. Resource management
7. Firmness and cooperativeness
8. Nonjudgmental approach
9. Skepticism
10. Technical realism
11. Analytical, methodical serendipity
12. Perspective switching
13. Fire in the belly
14. Resiliency
15. Technical aptitude (pp. 45–50)

As can be seen from the list, this is a complex set of skills for any one person to possess. Naturally, not every person who is hired as a systems librarian will have all of these qualities, at least at the beginning of his or her career. Many will receive training in systems-related work during the course of their professional library degree or at conferences, workshops, or related educational experiences after they have begun their jobs. The other important component of the education of a systems librarian will be on-the-job training, a valuable (but challenging) way to gain knowledge and experience.

Roles

Systems librarians may play various roles in a library organization, depending on the size of the library and the type of automation projects. A good source of the types of roles that a system librarian may encompass can be found by Wilson (1988):

1. Designer
2. Planner
3. Implementor
4. Consultant
5. Technology representative
6. Facilitator (pp. 43–44)

It may be obvious that a systems librarian may be a designer in that the new automation system is being developed for the library and its specific needs. One critical role for the systems librarian is to understand the needs of the library and staff in a new system and to provide input to the automation vendor or open source programmer on how best to meet those needs. Depending on what type of system that the library chooses, there may be limited or broad choices in the options that the library may select. A person who can translate the library's needs for a system into the options offered by a vendor will help ensure that the ultimate product is designed to serve the unique needs of that library.

The planner role is really the central core of the systems librarian job duties as the expectation is that this new position will be the staff person who serves as the primary project manager. This will involve planning the timelines, resources, staffing, budget, and all the other factors that are part of this operation. This planning will begin prior to the actual start of the project to determine if automation is the best way to meet the needs of the library all the way through the successful conclusion of the project if the library decides to go forward. It is absolutely critical that the systems librarian has excellent project management skills to accomplish this task.

The implementor role goes along with the planning role—one can have the most perfect plan in the entire library world but if that plan is never implemented, the planning exercises will have been in vain. The implementor role takes the plan that has been developed for the project and actually activates the various stages to make them real.

The consultant role is an ongoing responsibility of the systems librarian, not only for the discussed automation project, but also for any other technology projects that the library may be considering. It is important that the systems librarian be involved in these discussions, not only due to his or her expertise in various technology matters but also for his or her understanding of the technological environment of the library. Most technology projects in a library do not happen in a vacuum but become part of the overall technology environment in the library. Having the systems librarian's input into new projects will allow the library to make sure that each new project fits within the overall technology blueprint for library services.

Another ongoing role for the systems librarian is being a technology representative on behalf of the library. This role can be either internal or external. If internal, the role could be to keep senior management informed of the library's automation project; if external, it could be to represent the library at meetings with outside groups, other campus departments, library boards, etc. whom may need information about the library's technological needs and plans.

The facilitator role is another aspect of a systems librarian job that will carry over to all the technology projects that the library is planning. The critical part of this role is to serve as the facilitator between the staff (who want to add new library services) and the possible technological solutions for their desires. Quite often, the systems librarian will be asked to investigate how a new service may be offered and develop a solution that the staff can implement.

Organizational place

The systems librarian is quite often the hub of the library, as many of the current jobs cannot be done today without technology. This makes this position increasingly important as libraries move further away from print resources to more electronic resources and services.

The systems librarian position can be found in almost any department, depending on the size and type of library. For large public and academic libraries, the systems librarian is generally part of an overall systems department, responsible for all aspects of the library's technological footprint. In smaller libraries, the systems librarian may be part of either the public services or technical services, often depending on what library functions are being automated first. Although the position of a systems librarian is becoming more standard in most libraries, it is often an existing staff person who may have these duties added to his or her existing job description.

Whoever is acting in the role of the systems librarian must make sure that he or she has the support of the library administration for any projects and the necessary time and resources to be successful. Selecting and implementing an automated system is not something that can be done on a part-time basis. The amount of time and effort to achieve optimum results can often feel more like a full-time job.

Relationships

A good systems librarian must have superior "people" skills, as they will be constantly interacting with a wide range of actors during the course of an automation project (Figure 4.1).

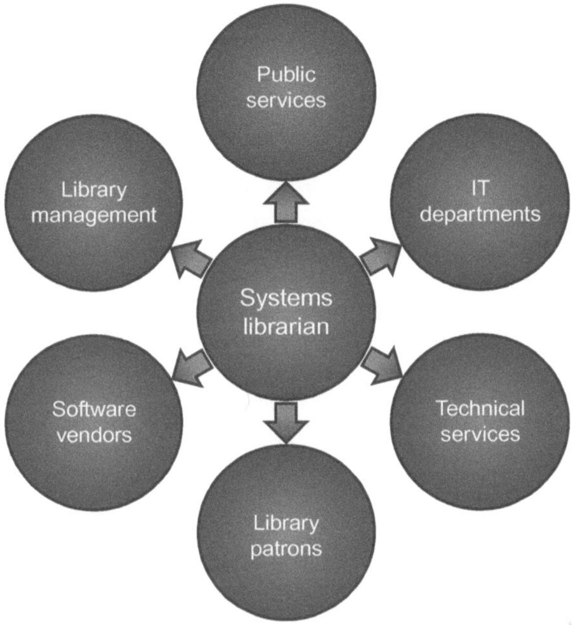

Figure 4.1 System librarian relationships.

The need for technical aptitude is important but equally important is the necessary social and management skills. It may be possible to purchase the best automation system on the market, but if the staff will not support it or the public will not use it, the money spent will have been wasted. It is absolutely critical that the people component—whether staff, faculty, public—is always part of the planning process.

Hiring technical staff

Depending on the type of library, there may already be established hiring processes for new positions. But hiring a new systems librarian may present some challenges if this is the first person with this title in the library. Below are some guidelines for the hiring process:

1. The first step will be to create a job description. Many librarians have written job descriptions for open positions or, if supervisors, have written them for other staff members. But, if unfamiliar with the job requirements for a systems librarian, the hiring manager should check some library listservs or library job web sites (like the ALA JobLIST) for examples of systems librarian jobs. After looking through the responsibilities, educational requirements, and skill sets that other libraries are requiring, the hiring manager may choose the ones that make the most sense for his or her type of library. At this point, it is important to ensure that the systems librarian job is focused on the systems work itself, especially if

the library is contemplating a major automation system implementation or upgrade. Often, new positions contain a wish list of all the duties that no one has had the time or energy to accomplish. An overly long or overly detailed job description will be confusing for the potential applicants, making it difficult for them to determine what the library is primarily seeking in a new staff person. It may also discourage other qualified candidates from applying.

2. One decision that will have to be made is whether a Master's degree in Library and Information Science (MLIS) for this position is necessary or whether an applicant with another type of background/education, such as computer science, would be suitable. The goal is to attract the best pool of candidates, so some flexibility in terms of degrees may be advisable.
3. Another decision will be what an appropriate salary (or salary range) would be for the position. Depending on the type of library, the pay schedule may be set by the city, state, or federal government. If the salary range is not set by the parent organization, a perusal of other job ads will provide a snapshot of the current library employment field and salaries. This information may be used as a comparison to other existing job classifications in the organization.
4. It is good practice to contact colleagues in other libraries who have a systems librarian and seek advice on what duties to include or exclude in a systems librarian position. Often, other librarians may have the best information on how a systems librarian position is created and how the position has changed or evolved since his or her first hiring.
5. The job advertisement for the new systems librarian should be posted in a variety of places to make sure that a broad audience of potential applicants is recruited. Depending on the type of library, there may be standard publications used for most job openings but it is also advisable to be open to posting the new position in a nontraditional publication or on web sites too. The successful candidate for this type of library position may come from an environment (e.g., a computer technology company) that may not be reached by the usual library placement services.
6. If possible, a committee should review the resumes of the applicants. It is especially important to involve the primary staff members who will be working with the new position and listen for their observations and concerns about the individuals under consideration.
7. Once the finalists have been selected, they should be invited to the library for an on-site interview. This is an opportunity to involve as many of the staff as possible in this process, as the systems librarian will be interacting with most of them over the course of any automation project. Often, scheduling individual and group conversations with each candidate will provide multiple perspectives on their qualifications and experiences.
8. Either before or after the on-site interviews, the references of each candidate should be consulted for verification of employment history and past work projects. Once this information is procured, it should be incorporated into the other comments that have been gathered from the staff during the on-site interview.
9. Once a candidate has been selected, he or she should be contacted with the job offer. Depending on who in the library does the hiring, this contact may be a senior staff person, the library director, or the head of the human resources department. At this time, the salary and job details are usually negotiated (which may be flexible or not) and the start date of the new staff member established.
10. Once an agreement with the preferred candidate has been reached, the other finalist candidates should be informed that a decision has been made.

If possible, the hiring process should not be rushed, as finding the right person will be one of the most critical decisions that the library will make. Having an experienced and knowledgeable person on the staff to help guide the automation project or implement new technologies will be worth the time and effort that has been put into the hiring process.

Human resources

As the person with the most responsibility for the technology projects in the library, the systems librarian is going to face many challenges dealing with the human resources side of any project. In large libraries, there may be a human resources department in charge of personnel issues but in many smaller libraries, the director, associate director, or department head often performs this role. Regardless of who is officially in charge of human resources, personnel issues will still impact the systems librarian.

Probably the most important element that the systems librarian will need to face in this area is the idea of change and how people react to it. The systems librarian will become the point person for introducing major changes to the library organization, and a complete understanding of this process and its effects is vital for success.

Another factor that the systems librarian may encounter is the issue of the past—that is, how have previous projects been implemented. There may be a level of distrust among the staff over the lack of success of past projects that may explain the current environment. Finding out this information and trying to allay these fears or apprehensions will be something that the systems librarian should be exploring.

Disruptions in the workplace, like the introduction of a new system, also can cause lingering problems lurking in the library background to come to the surface. The systems librarian needs to be aware that sometimes an obstacle in the current technology project is not actually about that project, but is a manifestation of an earlier problem that may never have been addressed or solved. The savvy systems librarian needs to differentiate between the "real" problems that pertain to the current technology project and what are slumbering problems that have suddenly come back to life. Enlisting other managers or department heads to help give guidance and support on these "other" issues may be critical to keep the current technology project from being swamped by past transgressions.

The advent of a new technology project may also impact the staff by hastening staff retirements or resignations by those staff who may not feel the energy or capability to succeed. This is not an unexpected occurrence as all organizations go through personnel changes but it may leave the systems librarian without key staff people with a long history of the library and its collections. This knowledge is often extremely useful in understanding past practices of library operations and why certain decisions were made as to data storage, record-keeping, or personnel. If a senior staff person decides to leave the library at the beginning of a technology project, the systems librarian should make a point to debrief the staff person and draw upon his or her long history of library operations before they depart.

To be successful, the systems librarian must have an extensive inventory of skills and play many different roles. Since an undertaking like the selection and implementation of a new integrated system represents a major undertaking by the library, it falls well within the scope of a project impacting the entire library staff. A wise systems librarian must have an understanding of basic project management concepts and use this skill set when developing the framework for their system implementation and/or migration project.

Project Management

5

Project management

Almost everything that happens in a library can be considered a project, whether it be implementing a new automation system, a new online catalog, or a facility remodel. One of the most important skill sets that any librarian should have is project management. This is especially true for systems librarians as the complexity of technology projects with their strict deadlines, complex data migrations, and multiple stakeholders make this a priority. Every project has a lifecycle with designated steps that can help guide both the experienced manager and the novice. Whatever the task, following these steps will help guide the project toward success (Figure 5.1).

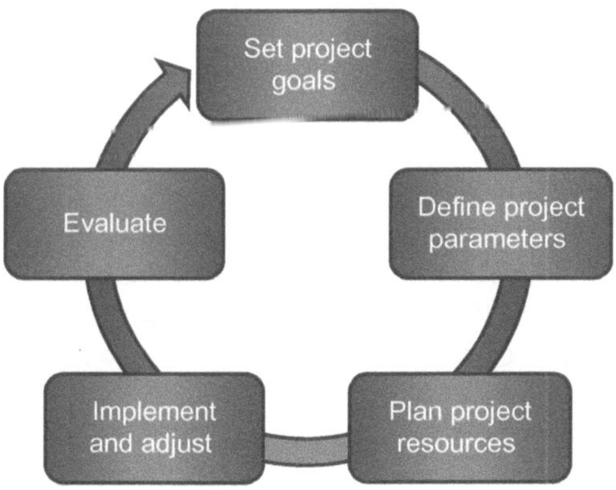

Figure 5.1 Project management life cycle.

Step 1: Set project goals

This first step may seem like an obvious first step but it is not always where many people begin a project. When confronted with a task, many people like to jump to the implementation stage right away, especially if they are facing a deadline. Beginning a project with immediate implementation may seem to be a way to ensure that the project gets completed quickly, especially if the project deadline is in the near future.

However, skipping this first critical step is probably the number one reason that many projects fail. The library cannot begin implementing a project until the ultimate goal of the project is known and understood. Beginning a project without a clear definition of the goal is like setting out on a car trip without a map. The car may be moving but the passengers don't know where they are going.

The first step should always be to define the goal. In some cases, the goal may be set by the library director, the board of trustees, or a direct supervisor, but at other times the goal may be the systems librarian to set.

An example of setting a goal for a new automation system would seem to simple. The goal may be defined as the implementation of the new system, but that really does not encompass the true meaning of the goal. It would be better to set the goal as the *successful* implementation of the new system with all staff fully trained and operational by a date certain. That goal is measurable (have all the staff been fully trained by the agreed date?) and accountable (is the new system fully operational?). Having a well-defined goal is absolutely necessary for a successful project and if the goal cannot be adequately defined at the beginning of the project, the project manager would be wise to wait until the goal is clear before proceeding.

Step 2: Define the project

Now that a goal has been set, it is important that the library define what this project is about. This is to serve two purposes, one external and one internal. Since a task group or committee does most projects, defining what the project is about is critical to making sure that everyone in the group is in agreement with the basic concepts of what the project entails. This will ensure that everyone is working together in a shared environment and understanding of the project before any work commences. This is the internal purpose of the project definition. The external purpose of the project definition is to make the project comprehensible to any outside group, staff members, library management, etc. who may need to be kept informed about the project and its status. It is rare that a library project operates in a vacuum; there are usually other projects and directives being implemented at the same time. Having a clear, jargon-free definition of the project will serve to remind all interested parties of what the library is hoping to accomplish.

Step 3: Planning—decide on team members and resources

Now that a goal and a definition of the project have been established, it is time to decide which staff members will be part of the project team. Using the example of the new automation system, one can assume that it will impact all departments in the library, both technical and public services. In this case, having a broad representation from all the library departments would be the best solution. It does not necessarily always have to be the unit or department head. Often a key staff member who may be in a subordinate position is in the best position to represent the interests of that subgroup.

The goal of team selection is to keep the team small enough to be manageable while at the same time being representative of as many library operations as possible. This way, the project team leader will hear all the various viewpoints as the project moves forward before they become problems.

Once the proper staff members have been identified, it should be made clear to them (and to their supervisors) that joining this team will be a major commitment of time and energy. Hopefully, there will be other staff members who can absorb some of their existing job duties to give them more flexibility in working on the project. If not, some of their existing work may need to be put on hold until the project is completed. This should be worked out in advance with the library administration so that the project team leader knows the priority that the library is assigning to this project.

In addition to team members, the library will also need to identify the resources that it will need for the project. This is the point where the project leader should begin to outline the budget that will be needed to meet the goals, a budget which has to be in place before the project can begin. Some possible items for the budget may include:

- Equipment that will be needed. Can everyone on the team use his or her existing computers, or will the library need to purchase more?
- Need for supplies. In addition to the regular office supplies that may be needed, are there any special supplies that are required for the project? For example, if a circulation system is to be included in the new automated system project, will the library need to purchase barcodes or security tags for circulation?
- Need for travel expenses. Quite often, it is advisable for someone from the library to attend library-related conferences or professional library workshops. This will provide the opportunity to network with colleagues who may be considering a new system or who have been through the experience already. Information that can be gathered from other automation projects may prove invaluable as the library begins planning its project.
- Need for additional space for the project team. Although this may not be an additional cost factor if the library has a large facility, identifying collaboration or workspace for the project team will be necessary before they can begin meeting.
- Need for hiring part-time or temporary staff to help with the project, especially if the library lacks the necessary personnel for the tasks to be done or if they cannot commit the necessary time due to other job duties. It may cost more to hire additional staff in the beginning (salary, benefits, equipment), but it may prove to be a worthwhile investment if it helps the project to be successful and finished on schedule.

Step 4: Planning—review the technology

At this stage, the technology review does not mean the decision on which automation system will be purchased, but the type of technology that will be used to manage the project. For example, will team members be communicating with group members via email or chat? Will documents for the project team be posted on a web site or an online cloud resource? At this stage, the technology has to be selected for internal communication as well as for how the project team will be communicating the project status to any necessary outside groups.

Step 5: Planning—write procedures

Another step that often gets skipped as time pressures mount and the project gets under way is the writing of procedures. Before beginning each project task, write out the procedures that will be followed and annotate them as necessary if the procedures change. This is important for several reasons. One, it will help clarify for the team members the project workflow and priorities. Two, it will be possible for the team to distribute the written procedures to various staff members working on the same task, knowing that there will be a consistent workflow based on the agreed-upon procedures. Three, it will allow team members to go back and troubleshoot problems (if they occur) by locating the step in the procedure where the error occurred. Procedures should be kept in a central location where everyone on the team has access to them and everyone has the ability to annotate them as the project evolves.

Step 6: Planning—develop schedules

Schedules may be imposed on the project team from outside sources or the team may be able to set its own, but there is a need to have a timeline for the project. There will be no way to measure if goals have been reached if the project team does not have some milestones built into the schedule to mark where the project should be by a certain deadline.

For example, the library director has told the project team that the current automation system has to be turned off at the end of the year due to the expiring contract with the old vendor. That date gives the team a final deadline for the completion of the project (the end of the calendar year); everything else must be computed backward from that date. The team may want to develop a schedule of 12 months with a deadline for each month (signing of contract, migration of data, etc.) as subgoals toward the larger goal of a successful system implementation. In this way, the team can measure progress each month to see whether they need to adjust staffing or workflows to meet the final deadline.

Step 7: Implementation and adjustment

Once the team has completed these six steps, they are now ready to begin the actual project. One thing to keep in mind is that a project is like a living organism: it is not static but is constantly shifting and evolving as the pieces are put in place. The team cannot always anticipate the problems that they may face when the project gets under way. Despite the best planning, there may be some unforeseen issues that force reevaluation of the planned approach. Very rarely does a project, especially one as complex as an automation project, go *exactly* as planned, but having this project plan in place will allow the team to decide how to handle the unexpected circumstances and what else in the plan needs to be shifted. However, if circumstances change so radically that the project is taking on a new direction, it is incumbent on the team leader to revise the project goals and definition to keep pace with the new reality or to begin a whole new project.

Step 8: Project evaluation

Now that the new system is being used for the library's daily work, it is important to set some time aside to evaluate the project. Although the project team may feel like they don't have the time to pause and reflect, the time they spend documenting the successes and challenges of the current project will be well worth their time, especially if the library is confronted with a similar project in the future. And, human nature being what it is, if the project team doesn't do the evaluation when everything is relatively fresh in their minds, it will probably never get completed.

Specifically, the library will want to review the process that it followed and document whether it worked well for the project or not. It will be project manager's job to pull the team back together to complete this important task.

This evaluation should be a written document, covering all aspects of the project from the initial discussions about the possibility of a new system right up through the final go-live date implementation. The library should solicit the feedback from the project team members of course, but also the input from other stakeholder groups—staff, patrons, library administrators, IT department—anyone who was impacted by the project and may have some insights to share. What the library should be trying to accomplish is a 360 degree view of the project, encompassing the aspects of what went well and those areas that could have been improved.

Soliciting this feedback can be challenging, but there are a number of ways to do it. The library can certainly hold either individual or a group meeting with the project team to solicit their feedback and ask them to follow up with a written description of how they feel that their area of responsibility succeeded or was unsuccessful during the course of the project. For staff members, a town hall type meeting for everyone to share their observations could be used along with some mechanism (perhaps a blog or web survey) where anonymous comments could be received. Some people might be reluctant to express their unhappiness in a group setting but would be willing to send in their comments if they are anonymous.

For soliciting patron feedback, the library could put a suggestion box on its web site for comments and/or a real box on the circulation desk for written comments. Again, the library should offer the option of anonymity for those who don't want to be identified as their comments are just as important as those who are willing to sign their names. What the library should be seeking is the broadest possible range of opinions and experiences so providing multiple avenues to receive this information is critical.

The library should be prepared in advance that not all the comments or feedback that it receives will be positive; in fact, much of what the library may encounter could be negative. These comments can represent a somewhat emotional response to the entire project as there may be many people in the library and patron community that are still in one of the turmoil stages of change. That does not mean that their current comments are not valid as the library is attempting to paint a portrait of the entire process. It is the library's job to wade through the feedback and decide which are the legitimate opinions that have merit as opposed to those that denote emotional venting.

The ultimate responsibility for the final project evaluation document should belong to the project team leader. Once sufficient feedback has been solicited, the project team leader should organize it to provide a history of the project, an overview of the process, and a list of successes and challenges. As the project team leader begins to prepare this document, there are some sample questions that may help provide a framework:

1. Was the project team composed of the right mix of members? Were there any groups not represented?
2. Was the communication plan used by the project team effective? What forms of communication worked best?
3. Was the timeline appropriate for the project? What areas fell behind and needed more time and attention?
4. Was the budget adequate for the project?
5. Was the staff training plan effective? Did staff members need more (or less) training in any specific area?
6. Was the workload evenly distributed among team members or did some team members have to take on additional duties?
7. Was the process open and transparent? Could it have been improved?
8. Was the team leader able to delegate work efficiently? Did they handle conflict in a constructive way?
9. Was the vendor cooperative during the process? How could the vendor–client relationship be improved?
10. Was the library able to achieve its goals? Was the final product what was anticipated during the planning process?

These are just some of the questions that the library may want to address as it sifts through the feedback that it received. As found in Cohn, Kelsey and Fiels (2001), the average length of time for an integrated library system to be used by a library was five years but it was far more likely to need a major upgrade within the span of three years. Since most libraries will be undergoing multiple projects over a span of years, this evaluation can serve as an important guide for any future projects.

Even the best project management skills will not provide a systems librarian with the knowledge to anticipate every issue that may arise during a system implementation. Quite often, there may be impediments to the project that may not be fully reflected in the project management life cycle. These hurdles are often due to reactions that the library staff may have to the idea of change and how they have experienced change in either their personal or professional lives. An awareness of the change process and the theories of change management will be invaluable for the systems librarian embarking on a major library project.

Change Management

6

An important skill set for the successful systems librarian is the ability to manage change. As the person who will, in many cases, be the point person for bringing wide-scale change to every part of the library, the systems librarian must be able to anticipate how change will impact the staff, public, or other stakeholders, and how to manage this change for maximum effectiveness. As has been previously stated, no matter how well-planned the library technology project may be, if the team cannot deal with the people part (of which change is the major component), they may face obstacles that will delay or derail the project. Change is one of the most misunderstood and, to many people, frightening, aspects of library projects. It is critical that the project team understand the process so that they can help others navigate these difficult transitions.

Definition

Before one can begin to talk about managing change, it is important to understand what it is and its properties. The classic definition of change usually involves a transformation or transition from one state to another. There is a change in seasons, a change of weather or even a change in altitude.

There is nothing inherently frightening in this definition; indeed, change is a relatively benign phenomenon that is found in nature. Why then do so many people react so negatively to change?

Looking at some more definitions may provide a clue as to the more negative implications of change. As is found in Toffler (1970): "Change is the process by which the future invades our lives" (p. 3). This definition of change might cause a bit more anxiety. For many people, the future can be an unsettling idea, especially when it comes to their workplace. One of the first things that many people will ask (whether out loud or silently) is how will this proposed change impact my job? Will I even have a job in this new future?

Change models

Is change just some random process like a lightning strike, or does it have properties and cycles that can be charted to help use as a guide?

Selecting and Implementing an Integrated Library System.

An examination of selected change models that have been profiled over the years by various social scientists and organizational behaviorists will provide more background. They will demonstrate that there are certain cycles that most researchers agree can be observed in any organization undergoing fundamental change (Figures 6.1–6.3).

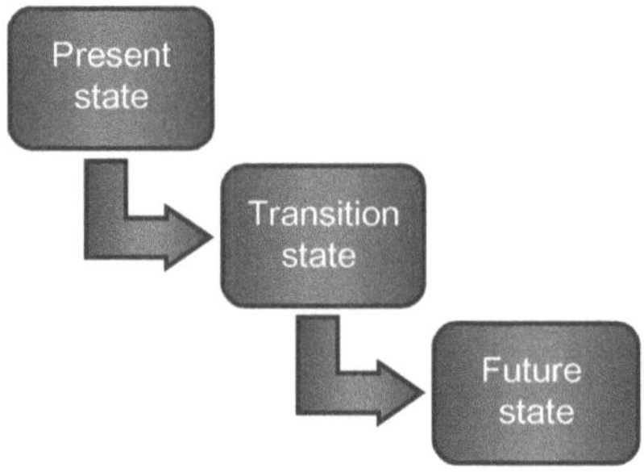

Figure 6.1 Richard Beckhard change model.

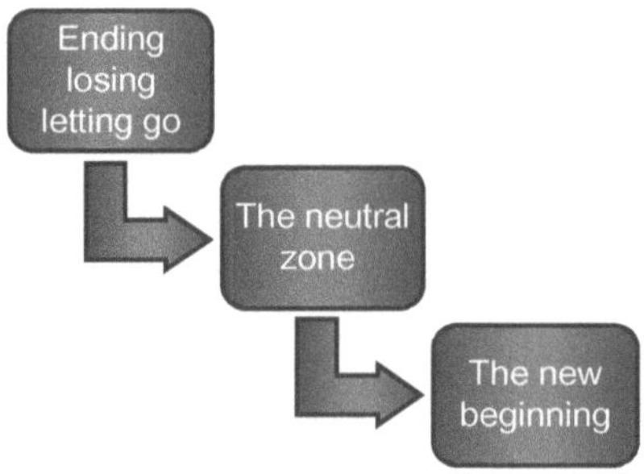

Figure 6.2 William Bridges change model.

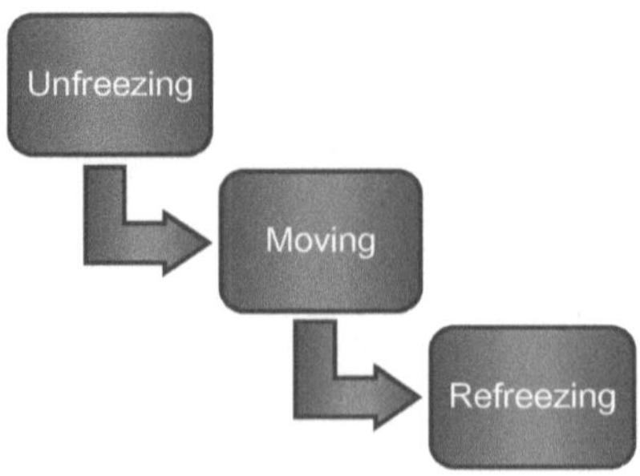

Figure 6.3 Kurt Lewin change model.

As is evident from the first three models, there is a definite similarity in the three-step process, even though the terminology is different. One begins with the present stage, goes through a middle stage where the actual change is occurring and concludes in a third stage which represents the new reality. What these three models do not detail however, at least by looking at the diagrams, is what is actually happening in this middle stage. This is the step that is of the most interest as that is the place where there is the greatest need for exercising management skills for a successful outcome.

A more detailed change model by Stephen G. Haines helps illuminate this middle stage. The Haines model provides more information by breaking down the middle stage into a series of stages. This change model can be used to highlight the stages in the example of implementing a new library automation system (Figure 6.4).

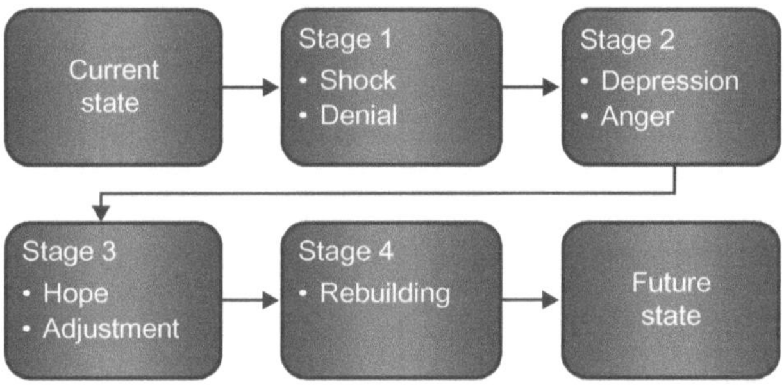

Figure 6.4 Stephen G. Haines change model.

The first stage in the Haines model is the impact stage, characterized by shock and denial. This is the stage where the staff may feel the first ripples of change in their daily work life, beyond the speculation and rumors. Up until the impact stage, certain staff members might have been able to ignore the proposed plans for automation and to believe that this project was just one more project that would never get funded or implemented. Suddenly, they start to see changes happening in the library—perhaps new equipment, a physical reorganization, and new workflows—that are impossible to ignore. This project is under way and the changes it will bring are real.

There are several ways that staff can cope with this stage. Hopefully, the project team has done a good job communicating with the staff up until this point to let them know what the project entails and how it will be implemented. They should be aware of the benefits to the change at this point and have a realistic picture of some of the challenges that they may face as the new system is implemented. Optimistically, most staff will either be in favor of the new project (or at least neutral) to give the team a reservoir of good will to draw upon when they hit the next stage.

The second stage is unsettled and difficult. This is the stage where morale is probably at its lowest point. People have begun to learn and use the new system but no one has mastered it as well as the old system. The feeling that staff members felt in their expertise in using the old system has disappeared and everyone, from senior staff to new hires, is at the beginning stage again. This loss of authority, especially for senior staff members, can be disorienting and can often be perceived as a loss of status as they are no longer the "experts" in their unit or department.

This is the most challenging stage for the team leader to maintain the momentum of the project and to keep morale as high as possible. This will be the stage where the change management coping strategies come into play. It can be a very emotional time too, as staff members lose their mooring posts and have to establish a new sense of identity under the new system.

The third stage of the Haines model is the hope and adjustment stage, where the staff starts to emerge from the depression and anger stage and begin the reconstruction of their work lives. The new system should be more familiar to them the longer that they use it, new procedures should be written, and life should be returning to normal in the library. This adjustment may take some time but with the proper training and support, all staff members should be feeling a growing sense of optimism.

The final stage in the Haines model is rebuilding. This is the stage where staff members have fully built a new work life, incorporating all the changes due to the implementation of the new system. A level of proficiency in the new system should be apparent, and all vestiges of the old system should be retired. The new system should be an integral part of the library operations, and any new services that were part of the system implementation should be in place.

Of course, what has been described is an ideal change process with well-defined stages, a system without flaws, and a staff willing to embrace the change. For those who have worked on library projects before, this is probably an unrealistic picture as any number of issues could surface before, during, and after the system implementation that could impact the outcome. Another point to keep in mind is that this change process is not a one-way street. Even those who reach the rebuilding stage may slip backward to the depression/anger stage if some part of the plan has setbacks. It is wise to keep in mind that everyone in your organization may be in a different place on this change cycle, and it would be rare to have everyone at the same stage at the same time. The members of the project team may naturally be impatient with the rest of the staff for what they perceive as their slow progress through the change stages, but patience is advised—the team members have had enough lead time to have already navigated through all the change stages. However, for many staff members, this may be a new process and one that will take some time to complete.

Support for change

Now that the team members and staff have a better conception of how change works, the staff can explore how to navigate change within the organization. The first point to remember is that change itself is neutral. It is the *reaction* to the change that makes the change process productive or difficult. The value that is assigned to the process, both individually and as a group, translates into how the process will be perceived and judged.

Naturally, the project team wants to have the best possible experience during any sort of change cycle, so they need to prepare themselves to help staff members approach this new project with a positive attitude. Understanding why people will support or not support change is critical at this stage.

As can be found by Hultman (1998), there several reasons why a group may be willing to support change:

- They believe that their needs are not being met currently.
- They believe that the change will make it easier for them to meet their needs.
- They believe the benefits outweigh the risks.
- They believe that the change process is being handled properly.
- They believe the change will work.
- They believe that those responsible for the change can be trusted. (p. 5)

Using the example of the automation project, these criteria provide a way for the library to approach change on a new level of understanding. One of the most evident reasons why people may want to support migration to a new system is that their current system no longer meets their needs. This may be due to many different factors—changes in formats being collected, changes in the scope of the collection, patron demands for new services—all of which may not be possible under their current system. By demonstrating to the staff that a new system will allow them to better meet the library's evolving mission, the staff should be ready to embrace the change process to meet their goals.

Another persuasive factor is to demonstrate to the staff that a new system will make it easier for them to fulfill their job requirements. Library workflows can be very detailed and often repetitive. Allowing some processes to be automated with a new system will free up staff time to focus on other library priorities. Since most libraries don't have the luxury of abundant staff (quite often the opposite), convincing them that a new system will be more efficient should help garner support.

Naturally, there will be risks with migrating to a new system, and the plan may encounter some disruptions along the way. The project team leader will want to be realistic about the challenges being faced, but by focusing on the benefits that the new system will bring to the library, staff will hopefully commit to sharing the risks for the overall benefits to come.

Trust is a central theme in the change process. Staff must be able to trust that the change process is being handled properly and fairly. This makes the need for clear communication, transparency in decision-making, and shared governance of the project critical. The project team will need to demonstrate that they are active listeners, open to input, and flexible in seeking opinions. This will create an environment of trust where staff will feel that they are vested in this project and that their input is being respected.

Of course, the staff will need to know that the change proposed will actually work as promised. In the automation project example, selecting a system that does not meet the needs of the library would be a catastrophic error. The project team will need to be realistic about what a new system can do and how this will impact the library operations. The team cannot promise to purchase a system that will address all the outstanding needs of the library unless they actually have the resources to deliver on that promise. Most system selection is usually based on compromise. One system, for example, may have a stronger circulation system than a serials system. Depending on the type of library, it would be wise to select the system that does well the most important function of the library, knowing that other parts of the system may not be as robust.

And finally, the staff must believe that those in charge of the change process can be trusted. Although the team may have done everything correctly to make the process open and transparent, there is still the issue of the personal integrity of those involved. The challenge for project team members will be to exhibit personal integrity in all staff interactions. This will be easier if team members have a long (and beneficial) history with the staff, but it can also be achieved if they are the new staff members hired specifically for this project. The new staff members must begin by treating everyone with respect and by getting to know the details of the library staff's specific workflows. Trust can be built in a fairly short time if team members are willing to be open and nonjudgmental about the current library operations and how they have evolved over time. It is very easy to squander the trust that has been carefully built with a dictatorial approach to the automation project, even though it may appear to be the easiest course of action. Building and maintaining the trust of colleagues will go a long way in making the changes the team is bringing to the library successful and accepted.

Before plunging into the selection and purchase of a new library system, some background work must be done to ensure that a new library system is necessary. Not every library problem can be addressed by technology, especially if the problem does not lend itself to a systems approach. Before embarking on a systems project, there has to be a demonstrated need for a new system. The first step in the decision on whether to purchase a new system should be a needs assessment to determine if there is sufficient demand to begin the system selection process. Once that has been determined, the range of available options in the library automation marketplace must be investigated as to the benefits and drawbacks for library system operations.

Needs Assessment and the Library Automation Marketplace

7

Needs assessment

As a library begins to contemplate an automation project, there are many steps that have to be done before one moves on to actually selecting a system. The very first step in any proposed automation project should be a needs assessment to determine if automation is actually the solution to the library's problems. Although many library operations can be made more efficient through automation, there may be times when another solution is a better choice. No one should purchase a new system simply to have one—it should be purchased and implemented only if it can help the library fulfill its mission.

A needs assessment is the perfect tool to help any library systematically examine its current environment and quantify the outstanding issues. Too often, decisions are made on purely anecdotal evidence that is not well researched or substantiated by facts. By using a needs assessment tool, the library can create a document that can help justify their need for a new automation system or address whether there are other possible options.

As its name implies, a needs assessment is a way of examining every staff position and workflow in depth. To begin, the project manager should set up some type of spreadsheet to collect the information that will be needed. For example, this is draft sheet for circulation activities (Figure 7.1).

Function	Who performs this task now?	Are other staff involved?	What equipment is used?	Are there any barriers with the current procedure?	How could this service be improved?
Creating patron records					
Printing hold slips					
Updating library hours					
Checking out books					
Reserving study rooms					
Collecting library fines					
Filling interlibrary loan requests					

Figure 7.1 Needs assessment worksheet.

The library can add as many functions or questions as necessary; the specific tasks and needs for each library may differ. The goal is to create a similar checklist for every department or unit, encompassing the major tasks that are currently being performed. This is not the time when solutions to the tasks that are being proposed; rather, this is a fact-finding assessment of what is currently happening in the library right now.

The best way to perform the needs assessment is to meet with every single staff person to go over the major functions of his or her job. In a small library, it may be possible for the systems librarian to do all the interviews but in a larger organization, this work may be divided among several staff members. The interviews should be conducted at the staff person's workspace when there are few distractions so that the actual functions can be demonstrated.

Once the project team has performed the interviews, they can compile the checklists into a single document that will provide a broad overview of the major tasks and how they are currently being performed. It is at this point that some analysis of the results can begin, focusing on what the staff members have identified as barriers to each procedure and how they might be improved. This is valuable information to have as the people actually doing the procedure will have the best idea of any problem areas and how they would like to see it improved.

After identifying the major barriers to service, the project team or leader should ask if automation may be a solution to some or all of the issues that have been identified. This may not be an easy process as some problems may readily lend themselves to an automated solution while others may not be so obvious. For example, if the project team identified that all patron records were being created by staff at the circulation desk, they may justifiably assume that an automated solution (creating them from a student database, for example) may be the obvious solution. But what about a problem like reserving library study rooms? Does the library really need to invest in an automated system to track these reservations or will a simple clipboard suffice? The answer may depend on the number of rooms available to be reserved, the presence of staff to help with the reservations at all times when the library is open, the ability to make keys available through a central secure locker, etc. Not every service barrier that is identified may be a candidate for automation, but it must be examined to determine its priority in library operations and its cost.

In addition to the individual staff interviews, the library may also be interested in collecting information from other stakeholders, such as patrons. This can be done by online surveys, focus groups, or information sessions to identify services that this population may want to see added or enhanced. This would also be a time when conferring with professional colleagues at other libraries may provide some background on how they are handling the procedures in their libraries. Researching professional literature will also provide examples of how other libraries are meeting their user needs.

Library automation marketplace

Assuming that the library has completed its needs assessment and decided that a new automated system is the right choice for the future, there are several important decisions to make. Will the library be choosing a system from one of the traditional library vendors, or will it be considering other options? There are definite advantages and disadvantages to each path and each one has to be thoroughly researched before moving forward on the project.

In the modern automation marketplace, there are many vendors offering a wide variety of services. There are vendors specializing in school libraries (Follett and Mandarin), public libraries (Biblionix and Polaris), academic libraries (Innovative Interfaces and Ex Libris), and special libraries (Lucidea and SirsiDynix). In addition to the traditional vendors in the integrated system service field, there are a few new competitors, OCLC and Serials Solutions, now selling their own integrated systems to customers who may have formerly only been familiar with their other electronic and bibliographic services.

The premier source of information about the automation marketplace is the annual Library Systems Report written by Marshall Breeding. This report is required reading for any system librarian wanting a snapshot of the current marketplace, complete with sales figures, individual company profiles, and business transitions. Breeding has been recognized as one of the leading figures in library automation. As a prolific author, his work is often cited for recognizing new developments in the field and current trends.

Software as a Service

Many vendors now offer a full range of options for libraries, from purchasing a completely profiled system hosted by the vendor (often called a turn-key system) to just the system software itself with minimal vendor support.

Software as a Service (SaaS) is a software licensing and delivery service in which the software is made available to the library on a subscription basis, rather than being purchased. It is analogous to the way that most libraries provide access to ejournals and ebooks: the libraries usually do not own these materials outright but provide access to them via a subscription service. Should the library stop paying the subscription fee, the eresources would disappear from its catalog, unlike the cancellation of a print subscription where the library retains the materials that it has already received and paid for over time.

SaaS is not really the same as buying a traditional vendor system. What the library obtains through the fee service is the software itself but it must supply the hardware that the software will need (unless cloud hosting is part of the SaaS package). However, similar to traditional vendor systems, the library will often not have access to the source code of the service and cannot modify it as desired.

The decision as to whether to use the SaaS option greatly depends on the individual libraries capabilities and infrastructure.

Cloud computing

Another concept that is often a part of the discussion of purchasing an integrated library system is cloud computing. This is becoming a relatively common way for organizations to manage their computing needs and many users are already using services that are based in the cloud without necessarily realizing it, that is, the Internet.

Cloud computing is the storage and access of data and programs over the Internet (called the cloud) instead of on an individual computer hard drive or a local server. In cloud computing, the servers storing the data and programs are often remote from the library location and managed by a third party (like Amazon Web Services) on behalf of the library or vendor.

There can be a number of advantages of cloud computing for a library system:

1. Cost. Large-scale cloud computing services may actually be less expensive for a library, as the cost of providing server space is spread among a large number of clients and not all borne by the library or the home institution IT department.
2. Support. Cloud computing services can often provide more efficient support in the event of a hardware failure. Since they are operating a large number of servers, they will often have spare parts stocked for repairs and will have staff on hand to respond to any emergencies, which may not always be the case with a server housed in the library or the home institution IT department.
3. Backup and disaster recovery. Cloud computing services will usually have servers in different geographical areas of the country (or world), which will allow them to quickly recover data should a disaster occur in any one location. This will make it more secure for the library to not rely on servers in only one location that may be impacted by whatever disaster has occurred in their specific location.
4. Storage capacity. With a cloud-based system, the ability to store unlimited amounts of data is available to most customers as the size and storage capacities of multiple servers are often greater than a local server. The cloud-based services can expand their storage capacity relatively easily, as they are constantly adding new customers while a local institution may have to wait to order parts or buy a new server to expand capacity.

But, there can also be some drawbacks to using a cloud-based solution. Some of these drawbacks would include:

1. Privacy. Because the library is turning over its data to a third-party vendor, there is an increased risk that this data may be compromised through negligence or error on the part of the cloud computing service. Especially for patron data, this potential data problem is a major consideration for the library system in terms of protecting patron rights.
2. Loss of local control. Another area that might be of concern to libraries with a cloud-based solution is the loss of local control, not only of the data but also of the hardware itself. The library may be forced to abide by rules or procedures that the cloud-based service has established for all of its clients, which may not be the best fit for the particular needs of the library. The library will have less direct control over its operations using a cloud-based solution and be subject to the priorities of an outside organization.
3. Staffing. By using a cloud-based service, a library may be looking to eliminate the need to have on-site IT staff for a local server and its operations. Although this may be possible to some degree, the absence of a dedicated IT staff member for library operations will leave the library without a person who understands the computing environment of the institution. In the event of an emergency or disaster, the library may have to depend on the cloud-based computing staff to remedy its problems, not all of who are going to be familiar with the library's needs and organizational structure. This could actually serve to delay the appropriate action while the cloud-based staff familiarizes themselves with the library's IT needs.

While the automation marketplace offers many options from traditional library vendors, there is another choice for libraries that is becoming an increasingly important part of the library system landscape. Open source systems offer many of the same features that the library can purchase from a traditional library vendor, without the cost of purchasing the software. Open source systems are free for any library to download but many libraries will find that choosing an open source system does indeed have many "costs" to consider. Deciding whether to opt for an open source system is something that libraries should seriously consider but only after investigating all the ramifications of such a choice.

Open Source

8

Open source

Most librarians are familiar with how to purchase products or services from a traditional library vendor. These vendors usually offer a product or service with a specific set of criteria (sometimes customizable) for a predetermined price. It is not much different than any other purchase that a library can make, be it for furniture, equipment, or library supplies. It is basically a transactional process by which the library "shops" in the automation marketplace, makes a choice of a system, and sends payment for the final purchase.

Open source, on the other hand, is a completely different sort of transaction. There are several competing products in the marketplace just as in the traditional automation vendor market to choose from but there is not the same selection and payment process as with a vendor. The attraction of open source is that it is free to the user; that is, there is no cost for the software but there are a host of other hidden costs that the library must be aware of before making a final decision.

An open source system is software where the source code is available to the general public at no charge. In a traditional vendor product, the library does not own the source code itself (which is owned by the vendor), but only the data in the system. If the library migrates to a new system, it needs to extract the data from the existing system but it does not get to keep a copy of the source code. With an open source system, the source code and the data are both the property of the library.

On the surface, this seems almost too good to be true. Communities of software programmers and librarians have come together to create an integrated library system, and are willing to share it with the world for no charge. As the cost of an automated system will be one of the largest expenditures in a library budget (at least for the initial software), the appeal of open source can be readily understood. But is free really free, or is it free like a "free kitten" is free?

Here's a quick comparison of some of the features of a traditional vendor system and an open source system (Figure 8.1).

Selecting and Implementing an Integrated Library System.

	Vendor system	Open source
Initial cost for software?	YES	NO
Ongoing maintenance fee?	YES	NO
Training materials supplied?	YES	NO
On-site training?	YES	NO
Customer support center?	YES	NO
Data migration support?	YES	NO
Contract needed for purchase?	YES	NO
System equipment support?	YES	NO

Figure 8.1 Traditional vendor system versus open source system.

Some of the above criteria may differ from vendor to vendor, but the basic outline is clear—when a library chooses an open source system, it will be relying much more on its own staff and expertise to implement an automated system than it would with a traditional library vendor. However, with the automation marketplace in flux, there are traditional vendors who are beginning to operate like an open source system while there are new players in the marketplace who will provide traditional services for open source systems. This may not make the job of a systems librarian any easier, although it will certainly increase the choices for any library considering automation.

Advantages

There are several advantages of choosing an open source system versus a traditional vendor system. The main advantage, at the outset, is the cost or rather the lack of cost. The software for an open source system is free—it can be downloaded at no charge and ready for use immediately. For a library with no initial budget to begin an automation project, this can be a way to procure a system without an initial outlay of funds.

An open source system is also the most customizable option that will be available in the marketplace. Because the library has access to the source code, it can make any changes that are desired to the code, creating exactly the system that best suits the needs of that particular library. It is entirely customizable to the library's *exact* specifications, something that is rare if purchasing a well-established vendor system.

Because the open source system has been developed by a number of programmers and librarians, the library will be able to take advantage of their research and development efforts. In theory, a system built for and by librarians should be more responsive to actual library needs, rather than one built by a vendor who may not truly understand the complexities of the library world. Being able to draw upon this expertise should provide the library with a system that has been tested and vetted by actual working libraries.

An open source system allows the library great flexibility in which library functions to implement in their library operations. For example, if a special collections library that does not circulate materials chooses an open source system, it may not want a circulation system at all, but it may be very interested in a descriptive cataloging system. However, many traditional systems will bundle these functions, making the library spend staff time implementing modules for which it has no use. With an open source system, the library can usually choose only those system components that are relevant for its institution, thereby saving staff time and resources.

The portability of an open source system is another advantage to many libraries. When shifting between proprietary systems, the staff has to learn and adjust to a whole new interface and system structure from the previous system. This can result in a loss of productivity along with low morale and frustration as the new system is being learned. In contrast, if a library chooses to switch between open source vendors, the company policies may change, but the basic library system will remain the same, making extensive staff retraining unnecessary.

Disadvantages

On the other side of the equation, there are some disadvantages to an open source system. One is cost. Although it is true that there will be no initial outlay for the software, there are many other costs that the library must bear that are ordinarily covered under a traditional vendor contract.
Some of these costs include:

- Hardware to use in downloading the open source software
- Programmer assistance to help customize the software to library specifications
- Staff to develop training materials
- Staff to do the on-site training
- Staff to handle all the data migration process

As is evident, there may be many additional responsibilities for the library staff to absorb with an open source system as well as the need for new equipment. In addition, if the library does not have someone on staff who is capable of doing the necessary programming to customize the software, it may need to add a new IT staff position or hire some temporary IT staff to make these changes. Placing additional burdens on the staff as they are already trying to cope with the implementation of a new system may prove to be challenging for morale.

Another disadvantage of using open source is that there will not be a customer service center or help desk to call when problems arise. Most open source systems do have active user communities, where the library staff can post questions. However, receiving assistance is dependent on the goodwill of participants and their ability to analyze the library's specific problems. Having a dedicated customer service desk that is tasked with responding to customer questions and staffed with system experts could be a quicker solution for any outstanding issues.

Because open source systems require a measure of IT support for server maintenance and possible programming changes, the library may need to determine how these services will be provided for their project. In many academic and governmental organizations, there is often a separate IT department to handle infrastructure and programming needs. It may be necessary for libraries with open source systems to collaborate with these outside units on their open source projects. This could lead to a loss of local control over the project, as it has to be fit within the timelines and workflow of another organization.

Open source vendors

To help meet some of the needs associated with open source projects, a new type of vendor has arrived on the automation scene to provide a bridge between open source software and libraries. These vendors work with open source software but provide to the library many of the services that a traditional library vendor provides with an established product. Two of the main companies who supply this market niche are ByWaters Solutions (bywatersolution.com) and LibLime (liblime.com), both of which support the Koha (koha.org) open source system.

Although the products vary between the two companies, each offers a range of services that include system implementation, data migration, training materials, on-site training, server hosting, and ongoing maintenance. In addition, each will provide the necessary customization of the open source software to meet the specific needs of the library, solving the need to seek programming assistance from another source that may not be familiar with the Koha software.

This range of services will create the same experience and environment of implementing a new system as that provided by a traditional library vendor. This option can allay staff apprehensions and hesitations about choosing an open source system, and would reduce the risks of not having the proper infrastructure for the automation project to succeed. However, this does not come without cost—although the Koha software is free, these other services will all be made available for a fee, negating perhaps some of the library's initial cost savings. Whether it is less expensive to go with these open source vendors versus a traditional library vendor will depend on the size, complexity, and services required by the library.

Tools

To help libraries determine if open source software is right for them, there is a terrific tool that is available from the Free Open Source Software for Libraries (FOSS4LIB) web site (foss4lib.org). This web site has a host of decision-making tools that will guide the library through the process of examining an open source solution.

These tools include:

> *Control versus responsibility: a survey tool.* Are you and your library more naturally inclined toward open source systems or proprietary systems? Do you want to host your systems yourself or have someone host it for you? Answering the 40 questions in this survey will help you decide.

Questions for a library's parent IT organization. Do you rely on your city, school district, college, or university for technology support? This tool is a series of questions that will want to ask that support organization to make sure you each understand the requirements for supporting open source software.

Costs of open source software. One of the strong virtues of open source software is that you don't pay licensing costs for the right to run the software. But there are other factors that need to be accounted for in deciding whether open source software makes financial sense for your library.

Software selection methodology. When you get to the stage of picking the software that is right for your library, you'll need a process for deciding among alternatives. This document describes the factors to consider when making that software selection decision.

Integrated library system (ILS) software selection methodology. Picking an integrated library system? This document guides you through questions and features to compare that are specific to ILSs.

Discovery layer software selection methodology. Discovery layers are becoming the common gateway for patrons to find library resources. Building on the general software selection methodology (above), this document focuses on discovery layer requirements.

As illustrated above, this is a comprehensive list of all the issues that a library should consider when making a decision about open source. Many of them are a series of questions that will guide the library through the process, helping it to consider all the ramifications of choosing this option. More importantly, this can be a way to gauge whether the library is really ready for this type of solution as it probes its organizational culture to determine the level of risk tolerance.

Faced with multiple decisions on which type of system would be best for an individual library, the systems librarian has a complicated task. To assist in mapping out the priority and progression of the selection process, a decision tree can be an effective management tool. For those system librarians who may be feeling overwhelmed by the number of decisions, this may also be a time to consider hiring a consultant. A knowledgeable consultant can provide expertise and guidance in navigating through this complex maze.

Decision Trees and Consultants

9

There are many factors that need to be considered as a library begins to think about an automation project. A useful tool to help organize preliminary options is a decision tree, a flowchart-like diagram that can help crystallize the options and possible consequences. Using this tool can help the library see the complete overview of the process and ensure that it is answering the basic questions first before moving on to the latter stages of the project.
Here is a sample decision tree for an integrated library system project:

Decision 1

After performing the needs assessment, the library must decide if there is sufficient need and interest in a new system to proceed. If there are other solutions to the needs assessment or lack of interest on the part of the library, the project should go no further (Figure 9.1).

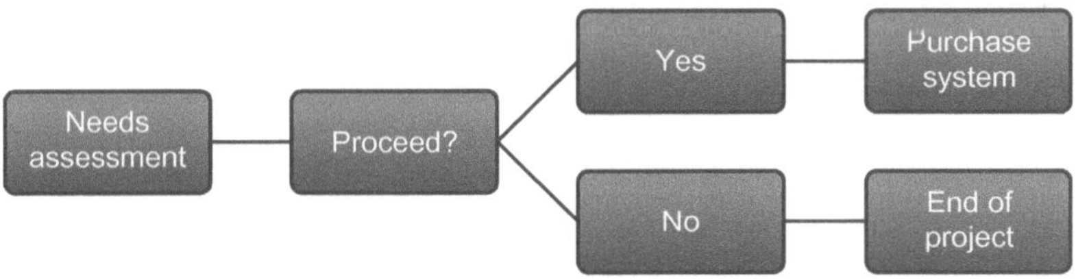

Figure 9.1 Decision tree decision no. 1.

Decision 2

The second decision (if the answer to decision no. 1 is yes) is to decide whether the library will be using a traditional library vendor for the integrated library system or going with an open source solution (Figure 9.2).

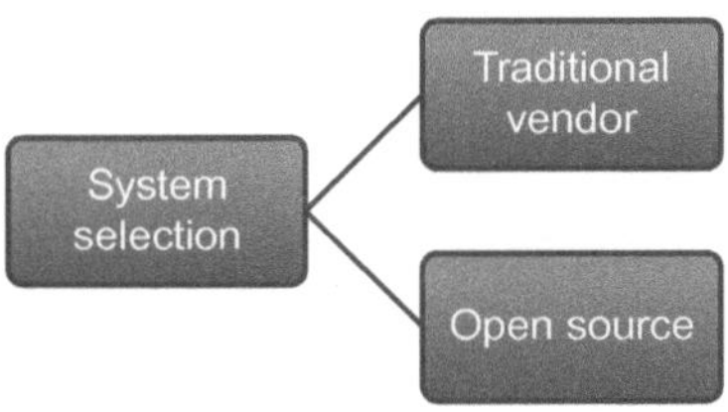

Figure 9.2 Decision tree decision no. 2.

Selecting and Implementing an Integrated Library System.

Decision 3

If the answer to decision no. 2 is to go with a traditional library system, the library will then have the sub-decision on whether to purchase a SaaS system from the vendor or a turn-key system where all software and hardware will be supplied by the vendor (Figure 9.3).

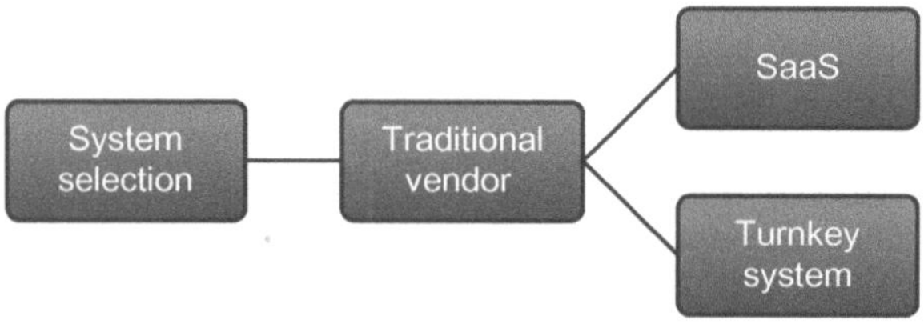

Figure 9.3 Decision tree decision no. 3.

If the option is to go with an open source system, the sub-decision is whether to download only the software or to engage a third-party library vendor who specializes in supporting open source installations (Figure 9.4).

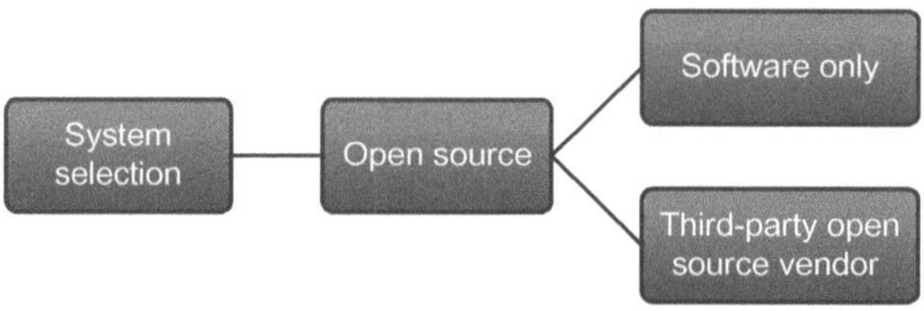

Figure 9.4 Decision tree decision no. 3a.

Decision 4

Whichever option is selected in decision no. 3, the library's next decision will focus on the equipment and access to the system. Will the library be using a cloud-based solution or will it be purchasing an in-house server to run the new system software?

Answering these questions in the right order is critical, as each decision builds on the previous decisions. Developing a decision tree can help clarify the options while providing a graphic representation to the project team as to the order and magnitude of its decisions (Figure 9.5).

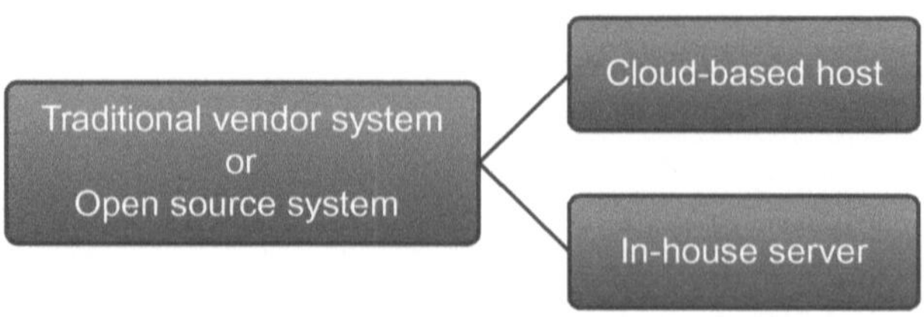

Figure 9.5 Decision tree decision no. 4.

Consultants

Another option to consider at this stage of the process is the possibility of hiring a consultant to help with the library's project. A consultant can often be a valuable addition to a project team, enabling the library to save time and money in the long run through their services.

One of the ways that a consultant may help save money is by his or her experience with vendors, contracts, and the general library automation marketplace. If this is the first time that a library has embarked on the selection of an automated system, it may find itself a bit overwhelmed with all the choices and decisions that it needs to make. An experienced consultant can help guide the library through the process, providing advice on which vendors to investigate and how to develop selection criteria.

Using a consultant can also help free up staff time by doing some of the work that would have to be done by others if they were part of the team. Having an extra person that can aid in the selection process can allow staff to continue other necessary tasks, like preparing the training materials or workflows that will be needed later in the automation process.

They may also provide the library with confidence that it has explored all the options for the project and considered the best solutions. An automation project is an enormous undertaking for any library. Having an outside perspective that provides evidence that all avenues have been explored before proceeding may prove to be very beneficial.

If a library decides to hire a consultant, it must be willing to do the necessary groundwork to make sure that the person hired is right for the library project. There are many sources for library consultants (e.g., libraryconsultants.org) containing a list of consultants, resumes, and expertise, or one can often find them through professional library organizations. Before interviewing a consultant, however, the project manager should do the following:

- Develop a written description of the library automation project and any relevant background.
- Develop a written job description of the consultant's job duties.
- Identify the timeline for when the consultant will be needed.
- Identify a budget and salary for reimbursing the consultant.
- Identify the outcomes that are expected from the consultant (reports, graphs, presentations, etc.)

Once these criteria have been established, the library should conduct the search as it would a search for a staff person in the library. It should solicit proposals from a variety of consultants to compare costs, agendas, and expertise. The library should check the references of the consultants to see how they performed at other libraries and whether the libraries would recommend them. Whenever possible, the library should bring them in for an on-site interview, involving the team leader and other members of the team who will be working closely with the consultant.

It may not be evident why a library would need to do such an extensive hiring process for someone who will only be employed on a temporary basis. There are two main reasons for treating this hiring decision as just as important as hiring a new staff member. First, the library will be entrusting this consultant with some vital components of the automation project. Depending on the range of services that the library has agreed that they will provide, they may be instrumental in helping the library select the system that it may be using for many years to come. This is not a decision that a library would want done by someone whose trust and ability did not inspire full confidence. Second, this hiring decision will represent one of the first management decisions of the project manager or systems librarian. Being able to hire a reliable and experienced consultant will demonstrate to the members of the project team and to the library administration that the project is in good hands. In many ways, the consultant will be the team's representative to the outside world and should perform their job with the same care and attention that other staff bring to their job duties.

Although consultants can serve a very useful purpose in the automation project, it is also necessary to be aware of a few pitfalls to hiring one. The first pitfall to avoid is delegating too much authority to the consultant. The consultant should not be making the final decisions on any aspects of the project—that is responsibility of the project team. The consultant is there to provide advice and guidance but should not be the ultimate decision-maker Also, the consultant will not have to live with the consequences of his or her decisions but the project team and library staff will, so all recommendations will need to be carefully vetted first.

Another pitfall is the temptation to use the consultant as a shortcut to not involving staff in the project. Although it may be easier for the team leader to seek advice from only one person, there are several risks with this approach. Not involving staff at this point in the project could mean a delay in their buy-in to the project, making it more difficult for the project team to solicit their support when the actual implementation begins. If key staff members are not involved, the process could also lose some very valuable perspectives and advice in the system selection process. Not every consultant will be familiar with all aspects of library operations and certainly no consultant will be as knowledgeable about library operations as the library's own staff.

With all this in mind, a consultant may still be a valuable option to consider. Although there will be a cost to hiring a consultant for the project, the advice and guidance that he or she may be able to provide may ultimately save time and money.

Once a library has decided to implement a new system, it must have a method to organize and quantify what type of system will meet its needs. The best method for doing this is to compile a request for proposal (RFP), a common tool used by many organizations to purchase large–scale items. The RFP can provide a library with a template for organizing the information about its current operations and future needs in one document which can be shared with a variety of vendors. Although time-consuming to write, the RFP can be an important instrument for forging staff consensus on the scope and options in a future library system.

Request for Proposal

10

The RFP process for selecting a new integrated library system is one of the most effective tools that a library can use. The RFP will help the library to organize its priorities, thoroughly document its needs, and establish an organizational framework for its project. Even if the institution does not require using an RFP, it is a tool that can help guarantee the success of a library systems project.

An RFP is a solicitation by an organization to outside vendors to bid on a service or commodity that the organization hopes to acquire. The details of the service/commodity are outlined (in detail) in the RFP itself, providing as much information as necessary to give the vendor the true scope of the project. The process is usually open to all relevant vendors in the marketplace for that service or commodity along with instructions and deadlines for submitting bids. Once the bids have been reviewed, the contract for the service/commodity is awarded to the vendor that best meets the needs of the organization (as outlined in the RFP) and often the vendor with the lowest price.

Developing an RFP can often be a time-consuming and expensive undertaking, especially in terms of staff time. Depending on the complexity of the organization, the RFP can be quite detailed and be several hundred pages long. Often, it requires multiple staff members to be involved as it may be a project that cuts across departmental lines and requires input from many sources. It is also not as quick as just signing a contract with a vendor because there is often a specified time period during which the RFP has to be public and another time period when the bids have to be reviewed.

Based on this information, why would an organization want to bother with the RFP process? One important reason is that the institution may *require* that all purchases for services and commodities go through an RFP process. This is particularly true for public and governmental entities on the local, state, and federal level. The RFP process is required to make the awarding of government contracts fair (to avoid corruption) and also to solicit the lowest bid for whatever commodity/service is being sought.

There are many other solid reasons for using the RFP process even if it is not required. In terms of the process of selecting a new integrated system, some of them are:

- It provides the library with a clear focus for the upcoming project. The RFP will force the library to define and prioritize exactly what it is looking for in a new system.
- It provides a clear way for the library to quantify its data and technical needs with facts and numbers, not anecdotal information.
- It provides a roadmap of the upcoming project for the library administration and staff.
- It provides an opportunity for staff to be involved with the project, solidifying their support for the system implementation.
- It provides a standard template to compare competing bids from different vendors.

Selecting and Implementing an Integrated Library System.

RFP process

Now that the advantages of using an RFP have been reviewed, how does a library go about creating one? These are the steps from the beginning of the process through the reviewing of the bids.

Step 1

The first step in the RFP process is to organize a team of staff members to aid in the creation of the document. The systems librarian (if that position exists in the library) would be the logical choice to chair a team for the RFP or the person with the lead responsibilities for the project. The other members of the team should represent the other units/departments that will be impacted by the implementation and migration to a new system. Since the goal is usually an *integrated* system, this should include a representative from all the various public and technical services areas. The library should also include a representative from the IT department for technical questions and someone from a management position to keep the library administration informed of the project's progress. It is important that the library select staff members who will be able to commit through the entire RFP process, as continuity in drafting and reviewing the RFP will keep the process on track and not lose valuable background information.

Step 2

Once the team is in place, the team needs to start defining exactly what it is looking for in a new system. The good news is that the team already has some of this data available to it through the needs assessment that the library did at the beginning of this automation process. This information can serve as a platform for the types of information that the team will need to gather in more detail as part of writing its RFP.

The approach that the team should take with writing an RFP is as follows:

1. Define what the library is seeking to purchase. Is it a turn-key system or SaaS? Is it an integrated system or a series of stand-alone modules? Is it to be an open source system?
2. Define the current library environment. What system (if any) is the library using right now?
3. Define the current library organization. How large is the collection? How many staff does the library employ? How many branches does it have? What is the patron population?
4. Define how vendors should respond to the RFP. Should they respond with a written proposal? Is email acceptable? What is the deadline for acceptance of bids?
5. Define the bid evaluation process. What factors will be considered in the final score? How will the evaluation be conducted? What is the time frame for a decision?
6. What is the library's timeline for implementation? Does the library want to implement all the modules at once or does it want them spread out over a set period of time?

A chart (Becker, 2009) can serve to assist with this process (Figure 10.1).

RFP element	Specific types of content	Purpose of information	Important considerations	Source of information
Introduction	Information about the library • size of collection • number of patrons • current systems Library's timeline • requirements for completion time	Provide clear picture of institutional environment	Needs to be current and accurate Needs to be realistic	Library statistics Mission/vision statement
Budget	How much have to spend • project as a whole • in phases (as applicable) • maintenance • training	Setting the boundaries for vendor Makes the library think about the process and how it will affect operations in the future	Understanding that it's not just a one-time expense Needs to be in context of other library spending What would be put off for another day if there's not enough money now? What is the funding source? (grant or on going budget line)	Library budget Library administration Other library's costs for similar projects Grant limits—may also include limitations on individual expenditures Other vendors
Vendor information	Financial information on vendor How the vendor should respond to the RFP	• Knowing if the vendor is going to be around a while • Inform the vendor • Comparing vendors • Setting guidelines for responses	Ability to compare vendors	Vendor Needs assessment Industry publications Conferences Other libraries
Systems requirement	Hardware/software requirements for new system (who will supply which?) Information about the current system Other systems in use (compatibility) Standards in use at library User and staff interface Hosting and support for network System availability How many people will be using the system		Specificity Accuracy Familiarity with types of technology (know what you're talking about) Don't needlessly limit (be specific, but open to other solutions) Group like-requirements together Have logical groupings Define level of importance Who will supply the hardware?	IT department Every other department Needs assessment -including users Other libraries with similar systems Other departments (transfer requirements)
Functional requirements	What do you want the system to do? -for users versus staff Number of users Expansion goals How do you want it to do it? What the interface should look like (look and feel) Problems to be solved	Make sure that the system can do things later	Getting input from all departments—what they really do How they would like to do it differently Include enough to compare vendors but not so specific that many couldn't meet them. Vendor might have a solution you haven't considered Required versus preferred	Needs assessment

Figure 10.1 RFP chart.

Point system/evaluation criteria	Breakdown of costs Provide the criteria and process Baseline requirements	Helps the vendor decide what the most important features are Establish minimums	Point system might affect how the vendor responds. Cross-purposes between library and vendor Need quantitative information Bound by criteria once it's issued	May get stuck with a bad contract if your criteria isn't clear or the right set for the project.
How vendor should respond	Deadline Contact information Format Electronic submission (how to access) Notification of intent to bid Response date (when decision will be made) Restrictions on communications Library and vendor liaisons Bidder information session Confidentiality		Need to be able to convert information into spreadsheet for evaluation.	

Figure 10.1 (Continued)

Step 3

Once the library has worked out the details of how the RFP is to be structured, it is time to begin a thorough documentation of all the processes that are to be included in the new system. Traditionally, this is broken down by the unit or departmental organization. An example might be:

1. Technical services
   a. Acquisitions
   b. Serials
   c. Binding
   d. Cataloging
   e. Electronic resources
2. Public services
   a. Resource sharing
   b. Circulation
   c. Discovery
3. Administration
   a. Invoice payment
   b. Budgets
   c. Reports

Under each of these headings and subheadings would be a specific set of requirements for that area, based on current library transactions and a projected amount for the future. The library will want to make sure that it purchases a system that will not only handle the current workloads but will have the capacity for growth as the library collections increase.

An example of an acquisitions section (Request for Proposal Shared Library Management Service, 2012) is outlined below:

> F.1.3. Acquisitions management
> F.1.3.1. In general, describe how the solution supports the acquisitions workflow, including, but not limited to, ordering, receiving, invoicing, claiming, and payment. Describe how order data is stored in relationship to bibliographic and item data, including, but not limited to, linking an order record to multiple bibliographic records.
> F.1.3.2. Describe the solution's support for automated selection, ordering, invoicing, and claiming, using standards like EDIFACT and X12. Can these transactions be completely automated? How is data sent and received in this manner integrated with acquisitions and financial modules? How does the solution check for duplicate records?
> F.1.3.3. Describe the solution's ability to import bibliographic records individually or in batches from a vendor, including, but not limited to, the automatic creation of order, invoice, and/or item records from data supplied by the individual institution.
> F.1.3.4. Describe your solution's support for ordering and claiming including, but not limited to, print and electronic submissions and what electronic submission protocols are supported.
> F.1.3.5. Describe how the solution supports the creation of brief bibliographic records for ordering purposes, if there is no bibliographic record available? Conversely, describe how the solution supports nonpurchased materials, such as gifts or government documents that require a bibliographic record but do not necessarily have an order or invoice?

The information in this section should be detailed and comprehensive. The library will want to provide the vendor with as much background information as possible as to its requirements so that they may have all the information that they need to respond. This is also the section where the library can describe how it would like to perform library functions in the future even if it does not have the capacity to do so at the present time. Many libraries are interested in migrating to a new system to gain new or enhanced functionality; detailing the library's expectations in this section will supply the vendor with the requirements that it is seeking in a new system.

Everything that the library is seeking in a new system has to be part of the RFP if it is expected to be part of the vendor response. Although it is often possible to add services that were not part of the original RFP at a later stage of negotiations, it may end up costing the library more if they are not included as part of the original response. The RFP process is intended to lead to a contract with a vendor and the library will want to ensure that it has done the proper homework on its end to lead to a successful conclusion.

Ideally, the library should be the organization in charge of preparing the RFP as they will be the most familiar with what is needed in a new system and how to organize it in language that will be readily understood by the vendors. Sometimes, however, another department in a large organization may be in charge of this process like a purchasing department. If this is the case, the library should try to get as many of its team members involved in the process of working with that third-party unit to detail exactly what is needed in a new system. Supplying them with a draft RFP or being involved in the discussions will help make sure that the RFP that is issued best represents the library's interests.

Step 4

Once the RFP is completed, it needs to be made public to provide the vendors with notice that the library is actively seeking a new system. Often, according to the policies of the organization, an RFP must be issued for a certain length of time, advertised in selected publications or web sites, with the same criteria being available to all. Because the library may have already done some preliminary investigations of different systems, libraries will often send the RFP to vendors that they would like to respond based on current products that they have in the marketplace.

Step 5

As the responses to the RFP start arriving, it is advisable to set up a spreadsheet (based on the RFP categories) so that the library may have an efficient way of organizing the responses for each vendor. This tool will help the library quickly scan the differences between the vendor responses as well as to identify any issue that the vendor neglected to include in their response. The goal is to compare the vendor responses to the exact criteria that were outlined in the RFP and how they rank with each other. This tool, as well as the original vendor responses, should be shared with all members of the system selection team.

Step 6

Using the evaluation method and criteria that the library outlined in the RFP, the selection team should begin ranking the vendor responses. Usually this process involves assigning a grade or percentage to each section of the RFP to arrive at a final grade for the entire vendor response.

For example (Request for Proposal Shared Library Management Service, 2012), see Figure 10.2.

Note that the library can include any categories that they want to score on this evaluation sheet but that they must correspond to a companion section in the RFP.

Step 7

At the end of the ranking discussion, the library should have identified the top candidate(s) for their new system. There will often be several vendors who rank close together as many of the new systems share many of the same features. It is at this point where the library may want to consider inviting the top-ranked vendors to the library for an on-site demonstration of their product to help provide more context to guide the decision-making. Most vendors will be eager to demonstrate their products and will often have a sample database for the library to use for the staff to experiment with different workflows.

These product demonstrations can be a very important method to help expose the staff to the potential new system(s) and build excitement about the changes coming to the library. It is critical that these demonstrations be scheduled at a time when the

Staff Functions and Discovery and User Experience
Scoring methodology for proposals that offer both staff and discovery functionality.

	Phase 2: Written Proposal	Phase 3: Product Demo	Total possible points	% of Total
Responsiveness to RFP	5		5	2.5
Collections and resource management	10	30	50	25
Description and metadata	10			
Circulation and resource sharing	10	20	30	15
Discovery and user experience	10	30	40	20
Systems	10	20	30	15
Price	30		30	15
Business references and vendor capacity	15		15	7.5
Total	**100**	**100**	**200**	**100**

Figure 10.2 Orbis-Cascade Alliance RFP grading tool.

majority of the staff can attend and that they be open to all members of the staff. Prepping the staff with some written documentation about the systems that they will be reviewing before the actual demonstrations will give them the background to fully understand what is being shown and how it might impact their job. A time for questions by the staff members should be planned into the presentations, along with the option for providing follow-up answers for those questions that needed further research.

Some libraries will rank the presentations too, just as they did the responses to the RFP (see the Orbis-Cascade example in Figure 10.2), while others may have a more informal method. In either case, the selection team should solicit feedback from everyone who attended the product demonstrations and from every staff member who has an opinion on what they liked or disliked about the systems under consideration. This feedback can be extremely important as the vendors are responding directly to the staff that will be using their system on a daily basis, often providing a different perspective than those of managers or department heads.

Step 8

At this point, the library should have quite a bit of information about the system(s) that it is considering. Hopefully, the list is growing shorter at each stage of this process to allow the library to focus on those systems that will best suit its needs. Another factor that the library may want to factor into its decision is the experience of comparable libraries that are actually using some of the systems under consideration. Quite often, the RFP will require that the vendor submit the names of other libraries using their product or services as references. Even if the library did not make this part of the RFP, it is advisable that the library solicit the feedback from colleagues as to their experiences, not only with the system but also with the vendor and system implementation. Although each system migration and implementation is different, it would be valuable to add the perspective of other libraries to the discussions, especially if they are of comparable size and complexity.

Step 9

Now is the time to make the final selection. Hopefully, the library will have reached a consensus within the selection team as to the best system for the library, or at least a majority opinion on what is the preferred system. A final report should be prepared for the library administration on the team's selection with supporting documentation from the RFP and the vendor's responses. It should be clear why the team has reached its conclusion, along with any potential limitations that the new system may not be able to address. If the team has done its job properly, the library administration should accept the team's recommendation and proceed to contract negotiations with the designated vendor.

Step 10

The actual contract negotiations with the vendor may not be the purview of the selection team. In larger organizations, the purchasing department may be the entity that actually negotiates contracts for all departments or it may be the library administrators that handle this part of the process. However, if the system librarian is in charge of the contract negotiations, it is strongly advised that the final contract be reviewed by legal counsel before it is finalized and signed. This is a binding contract on the organization, involving a great deal of time and expense, which will guide the library into the future. Ensuring that everything is spelled out exactly as desired and required in the RFP should be reflected in the final contract.

Step 11

Now that the final contract has been signed, this is the time to celebrate. Recognize the contribution of all the members of the selection team for their hard work and thank the staff for their involvement in the process. This is the time to start building excitement about the new system and enjoy the successful outcome to the RFP process.

Having completed the RFP and contract process is a major milestone in the project life cycle of a new integrated library system. Now that a new system has been selected and purchased, the implementation stage, where the actual work of setting up this new system for staff use starts to take place. One of the first tasks in the implementation phase is preparing existing library data to be migrated to the new system. If the library's data is already in electronic form, this migration process may be accomplished swiftly but many libraries are implementing a new online system for the first time. If this is the case, a retrospective conversion of existing data to an electronic format may be necessary.

Data Migration, Retrospective Conversion, and Barcodes

11

One of the more critical processes that will begin after the system selection will be the movement of data from the library's current system to the new system. Unless the library is planning to start with a blank slate, most libraries will have some data that they wish to retain and not have to reconstruct in a new system. Being able to migrate library data from one system to another will save a great deal of staff time and help the library make the transition to the new system quicker and more efficiently.

Depending on the library's current infrastructure, data can come from many sources. Some of them may be:

- Card catalogs
- Shelflists
- Online catalogs
- Serials checkin records (print or electronic)
- Patron records (print or electronic)
- Order records (print or electronic)
- Vendor files
- Accounting transactions
- Staff notes

The library may want to save as much of this information as possible and have it available for use in the new system or it may decide that some of it is no longer relevant. In either case, the library will need a system to determine which information is valuable for preservation and what is the best way to migrate it to the new system.

Step 1

The first step at this juncture should be to decide which data elements the library will be migrating to the new system and which will not be migrated. This is the time to cast a critical eye on library operations and workflows to see if there is redundancy in record-keeping and if data currently being collected has lost its value over the years. There is no reason to spend time profiling data for migration to the new system if the library no longer has a need for it.

The library will probably already have formed some opinions on this subject based on the work that the project team did in the needs assessment and RFP process. As team members observed workflows in the needs assessment and wrote up specifications in the RFP document, they may have come across procedures that made sense at one time but that have since lost their meaning. These are the very areas that the library should want to look closely at now to determine if the data that was being collected has any practical relevance for the future. If it does not, the library can skip writing up procedures for migrating this data and either archive the data that had been collected under the old system for historical purposes or discard it.

Step 2

Of that data that will be relevant for the new system, it will be necessary to assign a priority to the data sources to establish the project timeline for having this data converted. Because of the differences in formats and data complexity, the library may choose to prioritize the data that is easiest to migrate over that data that may take more time and effort.

An example of a library with both print and electronic data that it wishes to migrate is outlined below.

Under the hypothetical assumption that the only data that a library has in electronic form is the bibliographic record data and accounting transactions with all the rest of the public and technical services work still being maintained in paper files, this is what the data source table would look like (Table 11.1).

Table 11.1 Data source priority table (draft)

Data	Source	Format	Priority
Bibliographic record	Online catalog	Electronic	?
Order records	Order and vendor files	Paper	?
Serials records	Kardex	Paper	?
Patron records	Circulation files	Paper	?
Accounting transactions	Online system	Electronic	?

Assuming that the new system that the library has purchased is all electronic and that the library does not want to retain any paper files after the new system is fully implemented, how would it set priorities for the various data to be migrated?

One priority that would be high in any library would be the migration of the bibliographic records to form the new online catalog. Since the library's collection is not accessible without the catalog, this is a natural number one priority. It is also a smart decision to make this data migration the top priority as most library systems can very easily migrate bibliographic data from one system to another, especially if both library systems have been using standards, such as the MARC catalog record. This should be a relatively easy migration of data to accomplish with minimal staff time and preparation.

A secondary priority would be the patron records. Since patrons will not be able to borrow materials from the library without patron records, it is imperative that these records be migrated as soon as possible. Since the records that the library has right now are in a paper format, this will take a lot of staff time to input these manually into the new system. Because of this workload, it would be an excellent time to see if there is another source that can be used (campus registration files, voter registration files, online membership files, etc.) to create new patron records from an external source.

The third priority in the library example would probably be the order records, especially those records for which the orders are still active. Since these records represent materials that are currently waiting to be received, it is important that this group of orders be created in the new system as soon as possible. For those order records for which materials have already been received, this may be a perfect time to retain them in their original paper format for historical purposes but not spend the time and energy to convert them to the new system.

The fourth priority in the scenario would be to convert the serials checkin records to the electronic format in the new system. Hopefully, not all serials will be arriving on a daily basis, giving the staff time to convert these on a rolling basis as new materials are received and need to be recorded. It may be possible to assign subpriorities within this category to have those serials with a high frequency done first, with the annual publications or semiannual publications done at a later time.

The fifth priority would be the accounting transactions, not because they are not important but because they already exist in an online system. It may be possible for the library to continue to use its existing online system to perform accounting transactions while migrating the rest of the records to the new system. At that time, the accounting transactions (either just the active ones or all of them) could be migrated to the new system.

The revised chart would look like as given in Table 11.2.

This exercise now provides a path for the staff to follow in dealing with the data migration priorities and a blueprint to follow.

Table 11.2 Data source priority table (completed)

Data	Source	Format	Priority
Bibliographic record	Online catalog	Electronic	#1
Order records	Order and vendor files	Paper	#3
Serials records	Kardex	Paper	#4
Patron records	Circulation files	Paper	#2
Accounting transactions	Online system	Electronic	#5

Step 3

In consultation with the integrated library system vendor, the library will begin the data transfer process from the old system to the new one.

a. Create a test file of the library's existing records and load them into the new system
- Pull a representative sample of records from the library's database to test all formats of bibliographic examples.

b. Review the test files in the new system
- Check the test records for accuracy and review any problems with the vendor.
- Reload a new test file (if necessary) until all the records meet library specifications.

c. Send complete database record file to new system vendor
- Create a file of all records to be loaded into the new system and send it to the vendor.

d. Review database records
- Once the records are loaded, check them again to make sure that there are not any problems.

e. Create a "gap" record file
- Create a file of all records added or updated in the old system between the dates of the full database load and the current date.

f. Run statistics report
- Run any statistical, fund accounting, etc., reports on the old system for information that the library wishes to retain.

g. Print out record information
- Print out any information that cannot be automatically migrated to the new system and save for future input.

h. Cease maintenance in old system
- Stop creating or maintaining records in the old system and move all operations to new system.

Retrospective conversion

Another topic that often comes up at the time of data migration is retrospective conversion, the process where nonelectronic records are converted to electronic records. This term can apply to any record type but is most often used in conjunction with bibliographic records. Many libraries have had long-standing retrospective conversion projects over the years as technology shifted from the card catalog and shelflist to the online catalog.

When libraries think of this process, it becomes obvious why it often is associated with the implementation of a new library system. Since the library has already identified that the bibliographic record is the building block of the library collection, having them all available to staff for their work and patrons for their searching is a most desirable goal. Most libraries also want to avoid having both an online catalog and a card catalog at the same time because users of either one may miss out on the content available in the other format. As the online catalog and discovery layers become ubiquitous in the library world, the urge to convert all library records to an electronic format grows more urgent.

Since this is a process like many others, the steps that one should follow when contemplating a retrospective conversion of non-electronic bibliographic records to electronic records are outlined below.

Step 1

Under ideal conditions, the first step in this process should be to weed the library collection. There is no point in spending the time and effort to create online records for those materials that no longer fit within the library's collection development policy.

Removing these materials from the collection will ensure that the library is converting only those materials that are still relevant for its patrons.

Step 2

Once the weeding is complete, the library should do an inventory of the collection by comparing the shelflist to the actual materials on the shelves. There may be materials that have been lost or stolen over the years and are no longer part of the collection. There is no need to waste time and energy creating electronic records for materials that are missing from the library collection.

Step 3

Now that the library has an accurate count of the collection, it is time to analyze what it wants to have converted. Some criteria that the library may want to itemize include:

- Total number of records to be converted
- Number of languages used in the records
- Standards used (or ignored) in manual cataloging
- Material formats (books, DVDs, maps, etc.)

Step 4

This is the critical time to decide whether the retrospective conversion project will be done in-house by existing staff or contracted out to a library vendor that specializes in this type of work. There are advantages and disadvantages to both, which should be taken under advisement as the library considers its options.

Doing the project in-house may be advantageous for a small collection or those libraries without an urgent time frame. This would allow existing staff to work on converting records as they have time. In addition, the other advantages are:

- No initial financial outlay. The library will not have to come up with money for an outside vendor to complete the project but can use existing staff, salaries, and equipment to work on the project.
- Shelflist files will remain on-site. Quite often, an outside vendor will remove library files to their facility to work on the project. Doing the project in-house allows the files to remain in the library for easy access.
- Quality control. Since the existing staff and files will all be on-site, the ability to problem solve a bibliographic question about a record can be easily answered by a trip to the shelves to retrieve the material in question.

However, there can also be some serious disadvantages too:

- Workloads. Using existing staff to work on current cataloging and a retrospective conversion project at the same time may mean a loss of productivity for current materials due to conflicting time demands.
- The supervisory burden will increase for department heads or managers as they try to juggle both the daily workload and the demands of the retrospective conversion project.
- The timeline of the retrospective conversion project will probably be longer using existing staff and "spare" time.

On the opposite side, there are options for contracting the retrospective conversion project with an outside vendor. The usual method for these vendors is either to come to the library and make copies of the shelflist cards or to have the shelflist cards mailed to them for the duration of the project. They employ staff members dedicated to working on projects by searching a bibliographic utility (usually OCLC) for a matching record for the library client's shelflist card. Once they find a matching online record, they download the record to an electronic file, make whatever modifications the library has specified in its contract, and return this file of records to the library to use for loading into its new system.

As may be surmised, there are some distinct advantages to this approach:

- Minimal disruption. Since the work will be done off-site, there will be minimal disruption to the current daily workload of the majority of the staff.
- Timeline. Since the vendor has a staff dedicated to working on this project, the timeline for completion will be shorter than most libraries could do on their own.
- Quality control. The quality control of the project will be handled by the vendor per the library's specifications.
- Supervision. The vendor, as part of their ongoing business, will handle the supervision of the staff and training.

But there remain some disadvantages to consider too:

- Shelflist files. If the shelflist files are sent to the vendor, they may be off-site and inaccessible to library staff for the duration of the project.
- Cost. It may be necessary to pay the full or partial cost of the retrospective project when the contract is signed.
- Quality control. Since the vendor staff will not be as familiar with the library collection as the existing staff, it may be harder for them to find the exact match for a record from the shelflist cards.
- Communication. Since the vendor is off-site, it may be more difficult for library staff and vendor staff to communicate clearly at all times.

An important factor when contemplating the decision of whether an in-house or off-site project best meets the library's needs is the cost of an outside vendor. As Cohn, Kelsey, and Fiels (2001) found, costs can range from $0.25 to $6.00 per record; the average cost is $0.90 per record.

Some of the factors that will impact the cost are:

- Size of collection
- Publication dates (older materials often have no records for copy cataloging and must be input as original records)
- Languages
- Completeness of records being matched on bibliographic database
- How standardized the cataloging practices have been for the collection

Barcodes and RFID

In addition to the possibility of a retrospective conversion project, another decision that is often made at this time is whether to use barcodes or radio frequency identification (RFID) tags for library materials.

This most common type of barcodes is the one that is called generic or "dumb" barcode. These barcodes are purchased from a library supply vendor preprinted with consecutive barcode numbers, often with a specific library numerical prefix. They are called "dumb" because they have to be activated through a library process of linking; that is, applying the barcode to the library material and then wanding the barcode number into the material's item record. Through this manual process, the barcode becomes linked to that material and can be retrieved by that barcode number in the library's integrated system. This type of barcode tends to be fairly inexpensive to purchase but is fairly labor-intensive to install in a collection.

The other type of barcode is called a "smart" barcode and can be created by a library vendor from the existing data in a library's electronic files. Sending a file of electronic records to the barcode vendor for conversion begins this process. The vendor scans the files and creates a barcode for each individual title or copy in the file, linking the barcode number to the library record as part of the process. Once this process is complete, sheets of barcodes are printed (usually with the title, call number, or other identifying fields) along with the barcode number.

Since these barcodes are already linked to the library's records, it is imperative that the correct printed barcode is applied to the matching library material. Once the barcode has been applied, it is not necessary to do the wanding of the barcode number into the library record, hence the name "smart." These barcodes tend to be more expensive than the generic type and require some data profiling with a library vendor before they can be produced.

Depending on the size and complexity of the collection, the library may want to take advantage of the ability to produce smart barcodes for the collection. Although the initial cost will be higher, the money that is saved in terms of labor in linking the barcodes can be substantial. However, a caution for this type of barcode is in order too: if the smart barcode is placed on the wrong library material, it will not be properly linked in the library database, making circulation of the materials impossible. The library will need to make sure that whoever is applying the barcodes is a very detailed-oriented person who is not prone to mistakes.

Another choice for circulating library materials is RFID tags. An RFID tag consists of an antenna and tiny chip that stores bibliographic information about the item and a unique identification number. The tags are affixed to the inside cover of most library materials and, unlike barcodes, can also act as antitheft devices (libraries using barcodes have to use a third-party security device for the same results). Since each RFID tag contains the bibliographic information about the material to which they are affixed, they can also make performing inventory control a much easier task.

As might be supposed, the RFID tags are the most expensive options of the three, especially as compared to dumb barcodes. But unlike barcodes (either smart or dumb), the RFID tag is more than just a pointer into the library database. It can be used to track materials through complex library work processes and to easily perform inventory control. The choice of which option to select will be a matter of library budget, staffing, size of collection, and ongoing maintenance priorities.

Now that the library data is ready for the new system, it is time to prepare the staff to use the new system. Many contracts for a new library system will include staff training in the final negotiations or the training may be done by a third-party vendor or the library itself. However it is done, it is critical that the staff receive the best training available. This will help them use the system efficiently and prove to them that the disruption of migrating to a new system was worth the difficulty. This is also the time when training should be offered to patrons on the new system too, to prepare them for any new features that will be available when implementation is complete. While training is taking place and the system is being fully installed, there will be a continued need to troubleshoot any problems with the vendor that may occur on functionality, reliability, or data integrity.

Staff Training and Troubleshooting

12

Now that the technical details of the library's system project have been decided, it is time to focus on the groups of people who will be using the new system. As has been discussed in the previous chapter on change management, the new system that the library is introducing to the staff will be one of the biggest challenges that they will face in their library careers. It will change how they do their jobs, how they relate to their colleagues, and how they use technology in the workplace. It can be both exhilarating and exhausting, often at the same time.

At this stage of the implementation, the fact that a new system will be used for library operations should come as a surprise to no one. Although there may have been some staff members who would have preferred to ignore all the meetings, discussions, and demonstrations of new systems throughout the process, that is no longer an option. The need for everyone to be involved in the process is critical for its success: no one should have the option of opting out of the process no matter how many excuses are offered.

It is tempting as a project manager to bypass certain individuals in the library when putting together project teams as they have a reputation for being "difficult." By wanting to have the project go smoothly, it is often thought expedient to exclude those staff who might pose problems, hoping that they will just be swept along in the process with the rest of their colleagues. This is a completely understandable desire but one that is rarely ever advisable. A lack of involvement by just one person can have huge negative impacts on the total project.

There are several important reasons for needing to have everyone involved. First, if the library is planning for an introduction of a completely new system, it is likely that everyone's job in the library will be changing. Rarely is there a position in the library that will be isolated enough that the introduction of a new system for library operations will not have an effect on it. Even if, by chance, someone in the library is not directly impacted by the new system, they will still need to have a basic understanding of the new system to understand the work that their colleagues are doing and how the new system workflow manages library services.

Second, assuming that everyone's work will be changing, it is vital that all parts of the new system get implemented with the same level of efficiency and expertise as others. In a truly integrated system, there is no longer the luxury of one person performing work (creating records, posting invoices, etc.) that is restricted solely to themselves. The whole concept of an integrated system is that the work is shared by

Selecting and Implementing an Integrated Library System.

all. One person may create a record that can be viewed, edited, or modified by a host of other staff members. If one person is not conversant in how to use the system, they can disrupt workflows for the many staff members who are relying on them for their work.

Third, is the question of morale and fairness. The library has already asked many staff to be involved in this process, from the needs assessment to the RFP to the final system selection. The workload has hopefully been shared (to the extent possible) so that no one person has felt overwhelmed with the demands of this project. By allowing some staff members to not participate, the burden will fall more heavily on the other staff members who may resent their colleagues for their unwillingness to contribute. This feeling has the possibility of escalating into a serious morale crisis at the exact time that the library will need everyone to be fully engaged with the project.

What are the strategies for preventing this scenario? The most important strategy that the library can employ is to effectively communicate at every step of the process. It is true that one cannot overcommunicate. People absorb information in many different ways and at different times so it will be the project team's job to make sure that they use as many communication avenues as possible to reach the maximum number of staff. Not everyone is going to hear the information that the team is sharing at the same time with the same level of comprehension, which is why a variety of methods need to be employed.

Here are some of the communication options to consider when planning projects:

- Staff meetings—dedicate a staff meeting to the project as an introduction and provide regular updates on the project at all future meetings. If the library organization is too large for an all-staff meeting, make presentations at individual departmental meetings or functional staff groups.
- Email—send out regular updates on email or set up an email listserv for all staff.
- Blogs—create a blog for the project and encourage members of the project team to contribute updates, photos, etc.
- Social media—set up a Facebook or Twitter account for the project and encourage staff members to "like" or "follow."
- Web chats/conference calls—if the library has multiple branches in a large geographic area, set up web chats or conference calls with remote staff to provide information and answer questions.
- Videos—create video presentations for visual learners that provide project overview and context.
- Individual meetings—schedule individual meetings with key staff who might not be fully supportive or skeptical of the project.

The main concept to keep in mind is that whatever communication strategies are employed, they must be active; that is, they must "push" the information out to staff members. Waiting for staff to ask questions or to seek information on their own is not a winning strategy. Everyone is already overwhelmed with more information than they can easily absorb, so placing the burden on them to seek out additional information, no matter how important, is prone to failure. Providing clear and concise information in a variety of methods will ensure that the library's project will have a high visibility.

One of the most important methods for fostering involvement will be staff training. Although not everyone on the staff could be involved at the same level in the selection and implementation phase of the project (due to the time commitment and expertise of the project team), it is much more likely that everyone will be involved in staff training. How this important part of the project is conducted will be a key component of a successful project outcome.

Staff training

There are two general types of staff training that will probably be part of the library project: training provided by the vendor and training provided by the library.

Vendor training

It is quite common for training from the vendor to be a service that is available when purchasing a new system. Specifications for vendor training should be included as part of the RFP and subsequent contract negotiations, detailing how much training will be provided and what areas will be covered. Specifications should also be detailed as to the training documentation that will be provided (online, paper, etc.) and how it will be provided (via web conferences, on-site visits, etc.). If the training is not provided as a standard service by the vendor, it can often be purchased as a separate option from the vendor or from a third-party supplier that has been recommended by the vendor.

Training provided by the vendor has many advantages. The vendor trainers should be the most knowledgeable about the new system and have honed their training skills before other library organizations that have previously purchased their system. They should be able to share the latest information about the system and have current documentation to supplement the training. Vendor training is often done on-site (which is the preferred method as it is the most effective learning style), but for large installations, it may be done via web conferences.

There are some caveats that the library should be on the lookout for as it begins any training program supplied by the vendor:

- Trainers who are not prepared or knowledgeable.
- Trainers who do not tailor their training to the specific site but rely on a generic script.
- Trainers who are reluctant to answer questions or fail to follow up with answers after the sessions are complete.
- Trainers who are not professional trainers but are either technical people or sales staff used for training purposes.

Should the library encounter any of these problems (or others), it is incumbent on the project team leader to immediately relay his or her concerns to the vendor and seek a solution. Hoping that the training will get better if the library staff is disappointed with the first few sessions is a recipe for disaster. The harm done to staff morale in those initial sessions will be difficult to overcome. Any responsible vendor should be willing to work with the library to ensure that the training they provide will meet the specific needs of the library organization.

In-house training

No matter how much vendor training is supplied, most libraries will still need to do their own training to supplement it. It would be almost impossible for any vendor to cover every specific workflow situation in a complex library organization, much of which is based on local (and possibly nonstandard) practices. It is realistic to expect that additional training will be required for many aspects of the system and specific documentation prepared to support local practices.

The library's in-house training can start at any time; indeed, the library may want to do some basic overview training on the system before the official vendor-supplied training begins. What commonly occurs, however, is that the vendor-supplied training is done first and then any gaps in that training is done afterward to cover local issues. The in-house training can be as extensive or minimal as necessary but it does have to provide the staff with the necessary knowledge to perform their jobs and to use the new system efficiently.

There are several steps to set up an effective in-house training program. Although each institution will have specific needs, there are some general guidelines that should be followed:

- Select local staff as trainers who are good at training. This may seem obvious but it is not always put into practice when designing presentations. Not everyone has the skills needed for training. Some may be very enthusiastic or knowledgeable about the system but do they have the ability to relate that energy and knowledge in a way to reach a diverse audience? A good trainer is someone who is conversant in different learning styles, can explain concepts without using too much jargon, is comfortable in front of an audience, and is well prepared on the subject matter.
- Create documentation in advance of any class sessions and provide copies to all participants. This will allow staff to take notes during the presentations and to have a guide to follow when they return to their workspace to try out the new procedures.
- Leave time during each presentation for questions and comments. Although an outline of what is to be covered is essential, leave time to address any issues or concerns that come up during the course of the presentation. For those questions that come up for which a ready answer is not available, make sure to follow up with the group with more information as it becomes available.
- If at all possible, do the training in a space (like a computer lab) where each participant can either have their own computer or share one with their colleague. Hands-on training is the best way for staff to get comfortable with the new system and actually experiment with the various workflow options.
- If the library is using any presentation software (like PowerPoint), make the slides available to each participant so that they may review them on their own at a later time.
- Use current examples from library materials to illustrate staff workflows. Make sure to practice the presentation in advance to guarantee that the examples chosen effectively represent the workflows that are being demonstrated.

Patron training

Another important stakeholder group that cannot be forgotten is the patrons who will be using the discovery layer of the new system. Similar to the staff communication issues, the fact that the library will be moving to a new discovery layer should not be a surprise for the library patron community. Efforts should have been made from the contract phase of the project to keep patrons informed that a migration to a new system was being planned. This is the time to start educating patrons about the new system and how it will impact library services (hopefully for the better). This could be done through community meetings, handouts, posters, online videos, social media, or any other communication methods commonly employed by the library to reach patrons. The idea of this initial phase of communication is to let patrons know that a new system is coming and what they can expect in terms of new or enhanced services.

Once the system is implemented and ready for use, a training program needs to be established for interested patrons. An ideal way to gather information for what topics to include in patron training would be to hold a series of focus groups, working with small groups of actual patrons to discover what issues or challenges they encounter with the new system. Presenting the focus groups with some sample searches and watching how they navigate the new system will provide the library with valuable information as to what areas will require more training to be fully understood and utilized. Focus groups can also provide feedback on the design of the discovery layer web page, allowing the library to tweak the design of colors, fonts, and layout to maximize the user friendly design.

Once the library has gathered some initial information on the new patron interface from the focus groups, a training script and documentation should be prepared. Since the library probably has a patron community with different levels of computer familiarity, it may want to offer a series of classes ranging from basic users to advanced users, along with special collection searches that might be popular. For those patrons unable to attend an in-person class, the library may want to either record the class sessions or develop an online presentation that can be posted on the library's web page.

As part of the data migration phase of the project, patron data should have been migrated to the new system (if possible) to allow for circulation activities to commence when the new system goes live. This migration will allow the library to disable the old system to prevent patrons using an outdated patron record for checkout. However, many libraries choose to keep their public interfaces, old and new, running simultaneously for a period of time to ease the transition to a new system.

Is keeping both public interfaces (old and new) available for use a good idea or not? There are a number of distinct advantages and disadvantages of this approach: Some advantages to offering the old and new system as a choice:

- Can ease the transition to the new system.
- Can serve as a backup to the new system if any problems develop.
- Can be used as a teaching tool (e.g., here is how a search was done in the old system and here is how to do the same search in the new system).
- May prove helpful when troubleshooting data migration issues from the old system to the new.

But there are some disadvantages to offering the old and new system as a choice:

- May cause potential patron confusion.
- May delay patron adoption of the new interface.
- May be more costly (if the library has to pay the old vendor for the continued use of the previous system).
- May be more work for the staff if maintenance has to be done for both the old and new system.

The decision on whether to continue to offer the previous interface with the new one should be widely discussed in the library prior to the go-live date of the new system. If it is decided that the old interface will be available for patron use for a period of time, it should be made clear that this will be a temporary phase and advance notice has to be provided before removing the old interface completely from the library web page. Hopefully, the training sessions for the new interface will have helped overcome any patron resistance and the reliance on the old interface will no longer be necessary.

Troubleshooting

Now that the new system is live, the library will be encountering problems that probably are new to everyone associated with the project. This is not unexpected: no matter how thoroughly the project team tested the new system before going live, there is no way that they could anticipate every single workflow throughout the library. Some procedures (e.g., fiscal closing) are only done on an annual or semiannual basis and were probably not one of the new system features that were tested during the implementation phase. The library may also encounter some data migration issues in records that had not appeared before, as more staff and patrons begin to use the system and perform searches outside the sample searches that have been prepared.

Knowing that the period after the go-live date is likely to be a busy time for error reporting, it is critical that the project team has a plan in place to handle this influx. Although it may seem to the library that the bulk of the work has been completed with the successful selection and implementation of the new system, there is still much to be done for a completely satisfactory conclusion. This may be a time when the library might consider rotating some new members onto the project team to replace some members who may need a break after the implementation phase.

In this initial phase of actual system use, the library may want to create a "rapid response" team to handle any problems that occur. Remember, these problems being reported after the go-live date of the new system are either preventing staff from performing their jobs or preventing patrons from locating library materials. They must be considered of the highest priority, and every effort must be made to seek answers to them quickly.

In some problem cases, the answer may be as easy as a training issue—the system is actually working properly but the staff or patron did not understand how to use it. This can be quickly addressed by providing in-person assistance, revising documentation to highlight the issue, or pointing them to a frequently asked question (FAQ) site with answers to common problems. These sorts of problems should diminish over time as staff and patron become more familiar with the new system and more comfortable with the new options.

However, in many cases, problem reports will indicate that the system is actually not working properly—either it is not functioning according to the vendor documentation or there is an error with the actual library data. In these instances, the project team will have to rely on the vendor for assistance. Though these cases may not be as readily solved as the other cases, they are equally important and need to be addressed without delay.

As part of the contract negotiations, the library should have already worked out the parameters of reporting problems to the vendor and how they will respond. This is a standard part of most library contracts and many vendors have a well-established "help desk" as part of their customer service. Generally, when the library signs a contract with a vendor for a new system, it is assigned a specific project team that will be working on its implementation. Questions and concerns that arise during the implementation phase are routed through the library's designated implementation team where they are usually addressed quickly to make sure that the implementation calendar is not compromised. However, post-implementation, the library is usually moved to the general customer service help desk where the reported problems are one of many from all the other customers who are using the same system. This can lead to delays and frustration, particularly after the initial experience with speedy service from the library's designated implementation team.

Even with this shift to a broader customer service operation, the library can still ensure that its problems are addressed in the most expeditious way possible. Some strategies to employ are outlined below:

- If the problem reported to the project team is preventing a staff person from performing his or her job, it should be ranked as a high priority case. Most vendor customer service centers allow customers to prioritize their problem reports to differentiate those that are truly critical to ongoing operations from others that are less important. Use this designation for those problems that are impacting staff workflows and preventing people or groups from fulfilling their job duties but don't rate every problem reported as a high priority. The library will want the vendor to recognize that when it uses this designation, this problem really is a high priority and must be addressed immediately.
- Maintain a list of the problems reported to the vendor and make it available to the rest of the project team. The library can either have one person responsible for problem reporting or distribute the responsibility among various team members but it wants to prevent the same problem being reported multiple times. Maintaining a list of the reported problems and making it available to all (on a shared drive or a web page) will allow everyone to check whether a problem has been reported yet and what is the status of the resolution from the vendor.
- Monitor the vendor's web site for announcements of enhancements or system upgrades. Often, these announcements will be sent out to all customers on a regular basis but keeping abreast of the latest developments will aid the project team in responding to staff or patron questions about the system features/options.
- Follow-up with the vendor's customer service operations if the library's problems have not been addressed in a reasonable amount of time. Most vendors have a tracking system for customer problem reports, but staying up-to-date on outstanding problem reports should be part of the project team's responsibilities.

User groups

Another resource that can be of great assistance in troubleshooting the library's system problems are user groups. These groups, usually independent of the vendor, are composed of customers of the vendor who share information and provide peer assistance with system questions. In many cases, the user groups have set up listservs to which interested new customers can subscribe and ask questions, relying on those customers with more experience to help them find solutions to their problems. The larger user groups often have local or regional subgroups of customers and sponsor workshops or meetings to share workflow ideas. These groups are a rich source of information for new customers and often charge a nominal fee for membership.

Just as library organizations and services have been transformed, so too will library staff be transformed in the digital age. Many of the qualities that have made for outstanding library staff in the past will continue to be valued in the new era along with a host of new and innovative attributes. Recruiting and preparing staff to work in this environment will be an increasingly important function for all library managers. As technology advances into all aspects of library operations, staff that has the ability to grow and flourish in increasingly complex operations will be an important gauge of a library's success.

Staffing the Libraries of the Future

13

Just as library services have been transformed in the current library environment, so too will the skill set required to staff these new organizations. Although it may be easy to project that many of the skills and expertise that libraries will need for the future will be different from the current staff members, it is more difficult to know exactly *how* different. Will there still be a need for "traditional" library skills (cataloging, collection development, record management) in the future, or will these skills be obsolete? And, if it is true that these skills are no longer needed, what new requirements will have taken their place?

Before a discussion can begin about skills, a word must first be spent on the future of librarianship itself. Will there still be a need for professional librarians in the libraries of the future, or will the current staff positions be filled by another type of professional or paraprofessional? As has been documented (Griffiths & King, 2011), there will be a need in the future for librarians but with some important differences:

> *The future for qualified librarians is strong. There is a continuing need for librarians to reinforce and strengthen their roles as leaders, navigators, organisers, managers and interpreters of the expanding record of human activity and accomplishment in all its forms. However, the library profession and workforce face continued and increasing competition for qualified individual from the rapidly evolving information industry. Librarians need to recognize that, to remain relevant as society at large becomes much more aware of and engaged with recorded knowledge opportunities, they need to be proactive in monitoring and adapting to change, and to pay considerable attention to positioning and marketing themselves in this environment.* (p. 299).

The authors contend that there will still be a need for professional librarians but not necessarily those with the same skills that are currently employed. How then, to prepare a list of qualifications for a library future that contains so many different options? If one knew for certain how libraries will be used by various patron communities, it would be a relatively simple task to come up with a matching set of librarian skills to meet this future. Unfortunately, there is precious little agreement on what the future will look like and how it will impact staffing.

There are, however, some general observations that point to trends that will impact library staffs of the future:

- Libraries will operate with smaller staffs and leaner budgets.
- Libraries will rely more on technology and will need staff that are technologically literate.

- Libraries will increase their reliance on shared services (like shared technical services) to save costs.
- Libraries will devote more job positions to direct patron service for access to digital content.

For technical services staff, there are several changes that are already impacting their work and will continue to do so in the future. The continuing evolution of the MARC record for cataloging purposes has accelerated in recent years, in particular with the introduction of Resource Description and Access (RDA; http://www.loc.gov/aba/rda/) the new standard for resource description and access designed for the digital world. RDA was introduced in 2010 and was built on the foundations established by Anglo-American Cataloging Rules (AACR2), issued in 1978. RDA provides a comprehensive set of guidelines and instructions on resource description and access covering all types of content and media. One of the main goals for RDA is to create a single standard for cataloging all types of resources, both analog and digital, so that records for all materials will be compatible.

In addition, the Library of Congress has launched the Bibliographic Initiative Framework (BIBFRAME; http://www.loc.gov/bibframe/) to provide a descriptive framework for resources that appear on the web or in the networked world. This new method of description was developed to be able to cite library data in a way that differentiates the conceptual work (a title and author) from the physical details about that work's manifestation (page numbers, illustrations, etc.). It is intended to replace the MARC record at some point with a new structure that better meets the needs of digital items and their relationships.

One projection that can be guaranteed is that the library of the future will have a large component of technology. Not only will libraries continue to provide hardware and software in their physical locations for patron use but they will have to adapt to the ever-increasing reliance on personal devices and mobile applications. There will be many patrons who will continue to come to the library (both for using the collections and utilizing the space), but some of the most dedicated library users may never set foot within the building. It is these patrons, increasingly independent, that libraries will need to reach out to with appropriate technology.

It also seems clear that services provided to the patron community will need to change as a result of the upgrades to technology but also because of rising expectations. Patrons may no longer be content with borrowing a DVD from the library to watch a movie, but may expect the library to provide a streaming service similar to the commercial online providers.

This example provides a clear contrast of the staff qualities desired under each scenario. For the library purchasing a movie on DVD for circulation, most of the traditional library skills that are associated with library operations will suffice. Starting with the acquisitions and purchase of the materials, through cataloging and processing, the DVD does not represent a huge change from how books have been curated for many years. The loan rules may be different for the DVD and certainly the patron has to have a DVD player to watch the movie, but the basic library pipeline of acquiring and distributing materials would be recognizable.

In contrast, a library setting up a streaming service for movies will have different challenges. Since this service will most likely be a subscription service, the staff person responsible for acquisitions and procurement must understand the complexities of signing a contract with limitations and terms that do not apply to one-time purchases. Instead of putting the DVD on the shelves in the media collection, the streaming service will need to be added to the library web page by the IT department or webmaster, complete with instructions on how to use it. As for the reference and circulation librarians, they will need to have an understanding of copyright laws, rights management, and user restrictions to explain to patrons why they cannot download a copy of the movie to their personal device.

All in all, this is a much more complex transaction than purchasing a DVD, requiring a correspondingly complex set of skills. No matter what the future shape of the library, it is safe to assume that it will be a more challenging environment as new technologies and user demands proliferate. Libraries will need to adapt to these changes while not losing sight of their core missions, values, and patron goodwill.

Without knowing what new technologies and services will be in use, can libraries predict the types of staff they will need in the future? Although the future may be uncertain, it is possible to develop a list of qualities that will serve libraries well no matter what the future may bring.

In former days, the introduction of a new library service had a fairly well-defined path. When a new service was introduced or a new technology was added to the existing framework, it was usually followed by a period of stability before the next update or revision to that service. In the current environment, there will no longer be a period of interlude between major changes; rather, it is an ever-expanding cycle of constant change that no longer allows for the long periods of stability that used to be common. In particular, updates to software are being pushed out to users on a constant basis, often leading to the necessity of adopting new hardware or platforms to take advantage of the new features. Hiring staff who are comfortable in this new environment will be crucial (Figure 13.1).

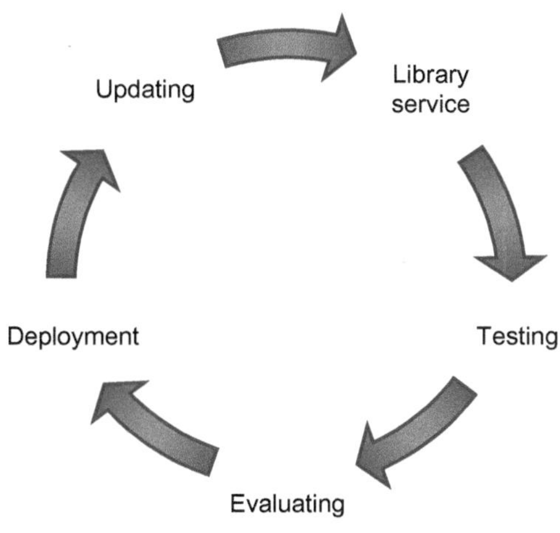

Figure 13.1 Model of constant change with new library services.

It is difficult to forecast the type of technical knowledge that future job seekers will need as some of the technology that they may be implementing is not even available yet. This is where the qualities that focus not on specific technical knowledge but rather the willingness to explore technology solutions with an open mind will be invaluable. Future librarians will need to have exposure to a wide variety of technologies, both in the technical services and public services sides of library operations, and be comfortable with migrating systems or implementing new ones as circumstances change. Keeping current with the library automation marketplace and the advances in social media will also be required to be effective proponents of new services to enhance library operations.

Knowing that the environment is shifting, the recruitment by libraries of new professionals to succeed in this atmosphere can be formidable. In addition to knowledge of programming languages, integrated library system experience, and graduate degrees, the library manager of the future must also seek some more intangible traits. In interviewing and hiring of new staff members, libraries must be aware of these qualities to seek out in potential job seekers:

1. Adaptability—The ability to change and adapt to circumstances as they evolve is critical in a constantly changing technical environment. The ability to be nimble when a shift in direction occurs is necessary for career longevity.
2. Inquisitiveness—The desire to be interested in exploring new ways of doing things is an ideal trait for the future. It is no longer acceptable to fall back on the canard that libraries will have to do things a certain way as "that is the way it has always been done around here." To always be questioning current practices ensures that unnecessary policies and procedures are not being maintained without purpose.
3. Irreverence—Coupled with the ability to be inquisitive, the ideal future candidates will also need the courage to challenge preconceived notions of how the library operates. Done in a respectful manner, the ability to seek justification for long-standing library activities can help the library examine whether some library folklore tales need to be retired.
4. Confidence—Future library candidates will have to not be afraid of trying new things even if they risk failure. Having a positive and confident attitude (although not reckless) will inspire others to be courageous too. Especially in a tumultuous time of change, having an upbeat attitude can help create an atmosphere of trust and support.
5. Collegiality—The need for collegiality is a core qualification. Just as almost everything done in libraries today is based on project management, most of these activities are also done by teams working together. The days of individual library employees working by themselves in separate offices is over; most of the work done today in libraries is collaborative. The need to get along with a wide variety of people from different backgrounds and ethnicities will be a foundational prerequisite for future library staff members.
6. Versatility—It is going to be challenging for future library hiring committees to write a job posting that adequately addresses the needs of any one position. Libraries are going to have to recruit people who are comfortable using a host of skills in their workplace, often a blend of traditional technical services and public services expertise. Based on lean budgets and expanded services, more cross-training between library staff members will be the new normal to meet patron demands within budgetary limitations. Based on the need for collegiality, the conventional pattern of hiring introverts for technical services and extroverts for public services is waning. Future candidates will need to have the same attention to detail as traditional library work to launch new services but also be able to interact with patrons

to help them use the new services most effectively. The fusion of both technical service and public service skills will be the hallmark of new library staff members.

While this list of qualifications may seem daunting (or impossible) to seek in any one individual, it can be a useful measure of what libraries need to cultivate in their new staff members. This is not to say, however, that all new staff must be hired to achieve the library's goals. Many current staff members in libraries have these traits already or are willing to learn them. Creating a new culture within an organization will take time and patience but having a roadmap of the type of staff members who are needed is an important first step.

The growth in the complexity and functionality of new library systems is just one aspect of the current transformation of the library world. Libraries are part of the larger societal changes taking place through the increasing proliferation of electronic access to information and constant communication. The digital landscape that has transformed how people interact with technology has had a profound effect on the role that libraries play in modern society. The rise in social media applications has allowed libraries to engage with their patron communities in new ways, both raising the awareness of the collections they offer and also raising expectations of new services that patrons may demand.

The Library Transformation in the Digital Age

14

There are as many scenarios as to the future of librarians as there are schools of thought on what libraries will look like in the future. In his book of essays (Janes, 2013), a variety of respected professionals were asked to come up with their idea of what the library would look like in 2020. The essays were provocative, engaging, and enlightening, pointing to a future with many different scenarios, all of them compelling. However, the array of futures that they imagined, with innovative services, modern technologies, and renewed mission statements, makes it almost impossible to generalize what the future will look like (except for the fact that libraries will indeed exist in 2020).

In the library scenarios of the future, it is true that libraries will have to change and grapple with the issues facing them. Already some of these issues are becoming part of the daily conversations in library circles and promise to accelerate as these trends continue. Some of the questions that libraries must ask of themselves are:

- With reduced or stagnant budgets, libraries will have to rely on more resource sharing. How will these partnerships be constructed? What will be unique about individual library contributions to a shared partnership?
- As libraries offer more electronic resources to their patrons, they will need to find a way to maintain patron attendance to avoid becoming a storage warehouse. Will libraries be stuck with an outmoded technology (books) as society moves on? Can the library function as the community portal for publishing, research, unique content, etc.?
- As libraries increase their emphasis on electronic and digital resources, what will be the impact on the idea of a local collection? Is there still a need for local copies of resources when information can be procured from all over the globe through downloads? Will the public still be willing to pay for libraries to house local copies when they can have an equal or better experience using their own technology?
- As technology becomes more ubiquitous, will the library still be relevant? Will the physical book collections become niche items for the privileged and not to the broad patron base needed for financial support? What will happen if the library chooses the wrong technology and ends up with the VHS tape of the future?

Library transformations

Perhaps to look into the future, an overview of the immediate past may shed light on where libraries may be headed. In the world that existed before the Internet, there were libraries; indeed, libraries have existed long before the concept of the Internet

Selecting and Implementing an Integrated Library System.

was invented. However, the birth and especially growth of the Internet in the past 25 years has had a huge impact on how libraries operate, store information, and provide access to their collections. Several distinct stages of library transformation can be described that mirror the transformations that have been happening to the Internet.

Library 1.0

When examining the history of libraries, the researcher will find it difficult to pinpoint exactly what is defined by Library 1.0 as very few people anticipated that there was to be a Library 2.0 or beyond. In a way, everything that came before the creation of the Internet can be considered Library 1.0, but the starting point can be narrowed to the period of the late ninteenth/early twentieth century when the model of the traditional library was customary. This library, which is familiar to most library patrons of a certain age, usually had primarily print collections, a classification scheme, open shelves, and the traditional public and technical services to maintain the collection and provide access. This model served in both academic and public libraries well into the 1970s.

At the same time that libraries began experimenting with electronic materials, the Internet was being made more widely available to the general public. And like libraries of that time, the characteristics of Web 1.0 reflect the beginning stages of the electronic revolution that was to occur. As found by Strickland (2014):

- Web 1.0 sites are static. They contain information that might be useful, but there's no reason for a visitor to return to the site later. An example might be a personal web page that gives information about the site's owner, but never changes…
- Web 1.0 sites aren't interactive. Visitors can only visit these sites; they can't impact or contribute to the sites…
- Web 1.0 applications are proprietary. Under the Web 1.0 philosophy, companies develop software applications that users can download, but they can't see how the application works or change it… (p. 2).

As can be seen, the analogy (although not exact) can be applied to the world of the traditional library as it has been identified. In the Library 1.0 world, the information was static to some degree. Although new titles were being added to the collections, the bulk of the collection did not change or was not updated very rapidly. Once a person had selected a book for reading or research, they often did not return to that same material on a regular basis for updated information.

There was also little room for interactivity in the traditional library. The library and staff made the selections for what materials to purchase, often with little or no input from their patrons. The interaction that the patrons could have with the library was through the reference and circulation functions, but this was often a one-way communication from librarian to patron.

Although the traditional library systems were not proprietary in the same way that Web 1.0 applications were (except for some of the automation systems), it can be argued that the access to the collections resembled a proprietary establishment. For example, patrons searching for materials in a card catalog (or early online catalogs) had to know the exact terminology that the catalogers had used when cataloging materials and could not use their own terms for searching. Indeed, it was not uncommon for libraries using the LC classification system to have the hardbound copies of the LC subject headings guide next to their card catalogs or terminals. These guides would serve to "translate" patron search terms into the often obtuse subject headings that had been established by the library community (Figure 14.1).

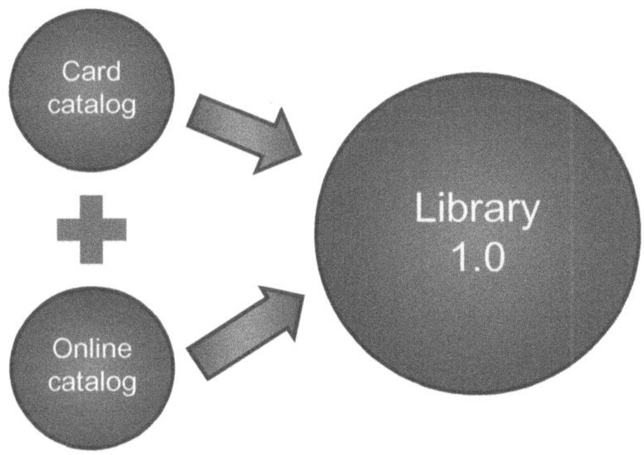

Figure 14.1 Library 1.0.

Library 2.0

Just as the web developed into a 2.0 version as it became more widely used, so too did libraries. This shift moved away from the types of traditional libraries that had been the standard into a new environment that encompassed major changes, especially in terms of patron services. As stated by Casey and Savastinuk (2007):

> *Every service we offer must work toward fulfilling the library's mission....For Library 2.0 to meet its goal of better library service, no service can continue forever without being reviewed to ensure it is still successfully serving library users. No service can be truly successful without some level of feedback or design assistance from the community the library serves.* (p. 7).

Library 2.0 is a transformation in the way library services are delivered to library users by:

- Providing new tools to make the library space (both virtual and physical) more interactive, collaborative, and driven by community needs.
- Encouraging collaborative two-way social interactions between library staff and library customers.
- Requiring user participation and feedback in the development and maintenance of library services.

The phrase "Library 2.0" first appeared in the blog *LibraryCrunch* (Casey, 2005), although the concepts that collectively make up Library 2.0 existed well before then. The term was a spinoff of "Web 2.0" (O'Reilly, 2005).

Probably the single largest evidence of change in the 2.0 world was the introduction of discovery platforms. This revolutionary tool shifted the emphasis from the card catalog and early online catalogs from strictly "searching" to "searching and retrieval" by embedding the options for accessing the materials as part of the record retrieval process. Yang and Wagner (2010) outlined some of the main characteristics of the Library 2.0 discovery world:

- Single point of entry for all library information
- State-of-the-art web interface
- Enriched content
- Faceted navigation
- Simple keyword search box on every page
- Relevancy
- Did you mean…?
- Recommendations/related materials
- User contribution
- RSS feeds
- Integration with social networks sites
- Persistent links (pp. 694–695)

In addition, discovery platforms often have more patron services than a traditional library offered, including allowing users the ability to tag items with their own subject headings, the ability to save searches, and the ability to suggest collection purchases.

Many of the Library 2.0 new features reside in the personalized library accounts that are part of this new stage. Often part of a discovery platform, these new library accounts have the option to store reading lists and notes as well as providing an overview of items checked out, fines owed, etc. Libraries can use the information in the library account to send personalized messages to each patron when their items are due for renewal or when new titles on a particular subject have become available (Figure 14.2).

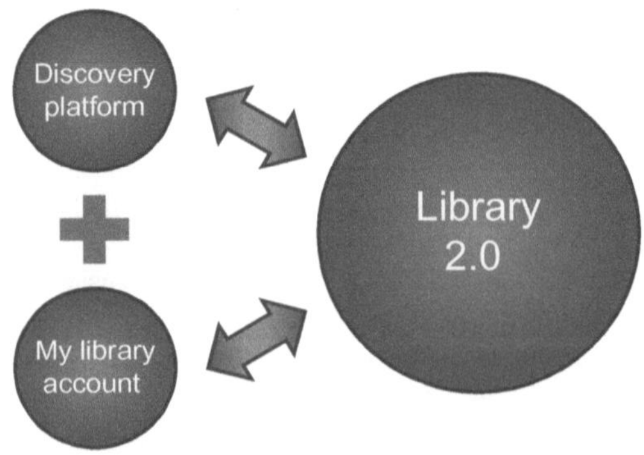

Figure 14.2 Library 2.0.

Library 3.0

The web world and the library world have now entered a new stage that promises even more changes to the way both the Internet and libraries are used. In this new Library 3.0 world, the emphasis on patron services has been expanded even further than in the 2.0 world. The emphasis at this stage is making the collections of the library available to any patron at any time, regardless of physical location. It is a much more proactive approach by libraries to delivering services and information to patrons, as libraries have belatedly recognized that there are competing services available to users for their information needs.

To compete in this new marketplace, libraries have begun to offer services that other companies have been offering to their customers for some time. The most obvious change can be found on the library search page. Instead of the multiple boxes for using Boolean searches in previous iterations of online catalogs, most discovery platforms use the single search box Google approach, the standard for web searches. Only in the advanced search will users find the multiple boxes and Boolean logic for more sophisticated searches.

There is also a shift in how libraries communicate with their patrons (now more commonly called customers by many institutions). Instead of waiting for patrons to contact them with requests for materials or questions, libraries are increasingly relying on social media to push information out to patrons and engaging in interactive conversations. Most libraries have a Facebook and Twitter presence, allowing patrons to "like" and "follow" the library for the latest news. Library blogs are popular, often with a comments section to encourage online discussion.

There is also a shift in the idea of the library as a space that is part of this transformation. In the modern library, the collections were seen as the primary asset of the library. The patron had to physically come to the actual library building to use the resources or seek assistance. As libraries progressed to Library 2.0, many of the services available to patrons became available online, especially the ability to ask reference questions, renew checked-out materials, or view their library account. However, patrons often still had to come to the library to use some of the print collections even as the electronic collections grew in size and scope.

In the Library 3.0 world, most library services are available to patrons online, no matter how far they are from the library's geographical location. As more library funds are devoted to electronic resources, the need for access to the print collections (some of which are replicated in electronic format) becomes less of a necessity. Although there will always be print materials that are not available in an electronic format, many of the popular materials requested by patrons have been digitized and made available online. It may be necessary for the serious researcher to come to a physical library for access to print collections but many patrons are finding that their information needs are being met through electronic databases, web resources, and eBooks.

As a result of this trend, many libraries are finding that their circulation statistics are lower than before as more people turn to electronic materials that do not register as circulating materials. At the same time, however, many libraries are seeing an increase in the number of patrons in their physical space although not perhaps for the traditional use of access to collections. The library is evolving as a meeting place for collaborative work with much more emphasis on fostering engagement among patrons rather than solitary study. Many libraries have reconfigured their physical spaces, storing more materials off-site to allow for an increase in user spaces. As book stacks disappear, study rooms with flexible walls, whiteboards, and projectors are taking their place, encouraging the collaborative process. In addition, cafes and snack bars have proliferated in library spaces, mimicking the atmosphere that many patrons find so appealing in their local coffee shops.

Communication tools have also been changing as libraries engage with their patrons. Most libraries now have a cell phone app for their patrons who use mobile devices to access their site and may use a short messaging service (SMS) to communicate via texting instead of email. The use of quick response (QR) codes to market library services has also become popular, allowing cell phone users to quickly scan the code for information on hours, computer availability, video tutorials, etc. In the place of printed maps, many libraries are also using geo-tagging to help guide patrons to where their requested item is located within the library building, along with maps to service points, rest rooms, or checkout stations (Figure 14.3).

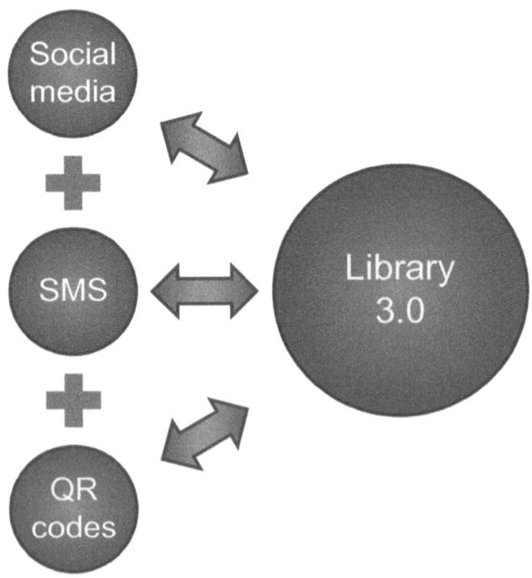

Figure 14.3 Library 3.0.

Undoubtedly, these transformations will continue to occur, both in the web world and the library world. It will be critical for libraries to stay abreast of the changes in the technology in order to retain and attract new patrons. As these shifts take place, the central mission of each library may need to be adjusted or revised to accurately reflect the needs of the community that it serves.

Social media

Part of the changing library landscape is the use of social media by libraries, a definite part of the Library 2.0 and Library 3.0 worlds. As ubiquitous as it has become among society as a whole, libraries have felt the need to use social media for their own needs, especially when it comes to patron services. Many libraries now have Facebook pages, Twitter accounts, and photo streams (like Flickr or Instagram), often set up by an enthusiastic staff person who felt that the library had to be part of the "social media" conversation. Often rolled out to the community with great fanfare, they sometimes end up with dated information or little interaction after an initial burst of activity.

This is the challenge of social media—not only to get it launched but to keep it relevant to the library's constantly changing patron population. Here are some questions to ask before plunging into the social media slipstream.

The first and most important question to ask is: Does this new service benefit library patrons? Unfortunately, this question is often overlooked in the rush to have the latest technology either because some staff member has read about it, heard about it, or knows of another library using it. There is also the desire to have the latest technology tool because it is new and shiny despite whether it actually fulfills a need at the library.

The second question that libraries should ask when considering social media is how much will it cost? Most of the social media platforms are free; that is, there is no charge to sign in and open an account. But similar to open source software, there is really no such thing as a "free" library operation. Someone on the staff will need to be responsible for the social media offerings if they are to be successful, potentially taking resources away from other library programs. Can the library absorb this cost and prioritize this service over other library services?

Another question to ask is what percentage of the library patron community will be served by this social media outreach? Will this service reach a large number of patrons or will it serve just a small percentage of the patron base? For some libraries, the cost of implementing and managing social media has to reach a substantial number of patrons to be worth the investment of staff time and resources. For other libraries, however, the cost and effort may be worthwhile to reach a smaller patron user group if that group is historically underserved by traditional library services. The other reason that a library may launch a new service without a defined point of return is to reach out to new patrons who may not be current users of the library but could become active through this new medium.

So, what are the steps to a successful social media program? Here are some basic concepts that could apply to almost any of the currently available options.

1. Before launching a new service, try to determine if there is a desire or need for this service through patron surveys, feedback buttons, focus groups, or observations from frontline staff.
2. If the library can determine that a need exists, it should be willing to experiment with new technologies but not be afraid to cancel them if they don't work out. It may take a few attempts before the library discovers the social media platform that works best for their patrons.

3. Assign dedicated staff to the initiative and make it part of their job responsibilities. Quite often, social media initiatives fail when the staff member who was doing it on a volunteer basis leaves and no one is assigned to carry it forward.
4. Do an assessment of the service after a period of time to see if it met its original goals (established by the library prior to the launch) and whether it is still meeting those goals. Is it worth the cost in staff time? Is there a new or better alternative on the market since the social media initiative was launched?

This is not to say that the libraries should wait until there is a clamor in the patron community for a new service before trying a new initiative. Libraries have sometimes been slow to respond to changes in communication styles; keeping up with how patrons communicate in other parts of their lives can help cement their relationship with the library. The library should be encouraging the staff to constantly be looking for methods to improve their ability to provide improved patron services by whatever methods are available. If there is a technology that is being widely adopted by other libraries or business entities, the library should do an investigation and consider a pilot project. It is important to know that not all pilot projects will succeed. Libraries must be willing to experiment while keeping in mind that a successful initiative will have to have the full backing of the library administration and dedicated resources to make it a priority.

Conclusion

The world of library automation is evolving, driven by the needs of libraries to offer new services to patrons and the continual advances in technology. Libraries are trying to meet these rising customer expectations while library budgets remain stagnant or decrease.

Libraries are also being impacted by twenty-first century social trends that may not be directly library-related but will have an effect on their patron populations. Some of these trends are:

Change in technologies

- The increase in the number of mobile devices, tablets, and smartphones shows no signs of diminishing. Information will have to be formatted to fit the display parameters of these devices in addition to standard desktop computers.

Changes in information formats

- As more patrons embrace electronic resources (audiobooks, eBooks, streaming media, digital materials) in their personal lives, libraries can anticipate that patrons will expect the same type of resources to be available in their local libraries.

Changes in publishing practices

- Amazon has had a major impact on the publishing industry. No longer just a seller of books, it is now a producer and publisher of content as well as the broker. Libraries will need to adjust their collection development policies to reflect this new reality as the publishing world becomes more centralized.

Changes in browsing habits

- The reduction in brick-and-mortar book stores has been one of the most dramatic impacts of the growth of online book sellers. As the ability of people to browse book store shelves has diminished, libraries may need to fill this gap for some segments of their patron communities.

Changes in reading habits

- The combination of changes in information formats and changes in electronic devices that people use to access information today has shifted reading habits. Many people are comfortable reading snippets of articles and news stories instead of searching for the full-text of an article (unless they are researchers). This "information-grazing" may lead to changes on how libraries display information to their patrons on their web sites or in their discovery layers.

These various trends, along with many other throughout society, will change how patrons view and use libraries and the place that they will play in the public sphere. However, the delivery of information (in whatever format) will undoubtedly remain one of central tenets of library services. Choosing a library system that allows the library to meet their mission and goals will be a critical component for success. By carefully selecting a system that meets the library's current needs and yet is flexible enough for the future, the library will ensure that it remains relevant in the digital age.

Appendix: The Orbis-Cascade Project

Impact of technology

It may seem obvious from the discussion of the ongoing library/web transitions, but technology—its growth, value, and ubiquity—is going to be the driving force in the future of library organizations. The traditional integrated library systems that have been in use for the past 25 years are in the process of being redefined for the future, propelled by both library demands and the application of new technology models. How this will impact libraries is already being seen in some new projects currently under way.

Orbis-Cascade project

One of the most exciting projects involving a new approach to library systems is being conducted by the Orbis-Cascade Alliance, a consortium of 37 public colleges and universities in the Pacific Northwest. Having long enjoyed a system that enabled direct patron-initiated library borrowing between the campuses, the Alliance began looking for a shared integrated library system in 2011 to replace the 37 individual Innovative Interfaces systems in use by member libraries. Following a successful request for information (RFI) proposal, the Alliance drafted an RFP that was issued to the automation marketplace in January 2012. (For a full overview of this part of the project, see Cornish, Jost, & Arch, 2013.)

What was unique about the Orbis-Cascade RFP was the stated intention to move beyond the current integrated library system model to a new model with one shared system for all member libraries. As found in the Orbis-Cascade Alliance Request for Proposal Shared Library Management Service (2012):

> *Through strategic planning, the Alliance has concluded that improved services, cost efficiency, and prospects for cooperative collection development and collaborative approaches to technical services will be facilitated by a move from more than 30 stand-alone Integrated Library Systems (ILS) to a single shared solution. More than simply replicating legacy systems in a shared or cloud-computing environment, we are issuing this RFP with the expectation of contracting for a unified set of services that exceed traditional ILS capabilities.* (p. 3).

While it is true that many library consortia have shared a system for the "front end" of discovery and retrieval, the Orbis-Cascade Alliance was specifically seeking a system for sharing both the public discovery search and retrieval function with the "back-end" functions of technical services and circulation. This would be a true "shared" system where all parts of what were stand-alone institutional-specific systems were now to be merged.

While seeking cost savings through the sharing of a single system was one of the primary motivations for this RFP, there was also the desire among member libraries to leverage their collections and services for maximum efficiency. The realization that no library enjoys the luxury of ever-expanding budgets to purchase everything that their patron community needs or wants makes reliance on library partners critical to fill the collection gaps. Only by pooling resources will libraries be able to expand their collection scope through collaborative collection development and cooperative purchasing.

Most of the automation vendors who expressed interest in this project (through their responses to the RFI) submitted responses to the RFP. After several rounds of product demonstrations and negotiations, Ex Libris was chosen in June 2012 as the vendor for the new system. The selection of Ex Libris was based on numerous criteria, especially their understanding of the shared system concept that Orbis-Cascade was hoping to implement. The final decision to move forward was made by the Orbis-Cascade Council, composed of the deans and directors of the 37 library institutions.

The Ex Libris system represented a major change in the way that the member libraries had formerly operated, as a shared system necessitates the use of shared policies and procedures. Now, all library workflows, not just the search and delivery functions, had to be coordinated with other libraries and workflows adjusted according to Alliance-wide policies. This extensive collaboration meant that a new governing structure for the Alliance had to be established with numerous working groups to create and disseminate the shared policies.

The Alliance formed the Shared ILS Team in September 2012 to manage the transition to the new Ex Libris system. The team was composed of a program manager (a new Alliance staff position), an implementation team, as well as the chairs of the functional working groups that were established to develop policies and procedures. Members of these functional working groups were drawn from volunteers across all the member libraries (Figure A.1).

The Orbis-Cascade project divided up the 37 libraries into four cohorts, with a mix of large and small institutions in each group. Beginning in June 2012, the first cohort went live with the new system, followed at 6-month intervals by subsequent cohorts. The total migration of all institutions was to be completed by January 2015.

The complexity of this project was reflected in the number of teams and volunteers that were critical to begin the implementation phase. It quickly became apparent that there were major differences between the needs of the largest library system, the University of Washington (UW) and the smaller members of the Alliance. The number of resources, patrons, and languages in the libraries of the UW system quickly uncovered limitations of the new system that may not have been apparent or important to the smaller institutions. Since the UW was in the first cohort to go live, many of these problems surfaced immediately and led to a difficult transition.

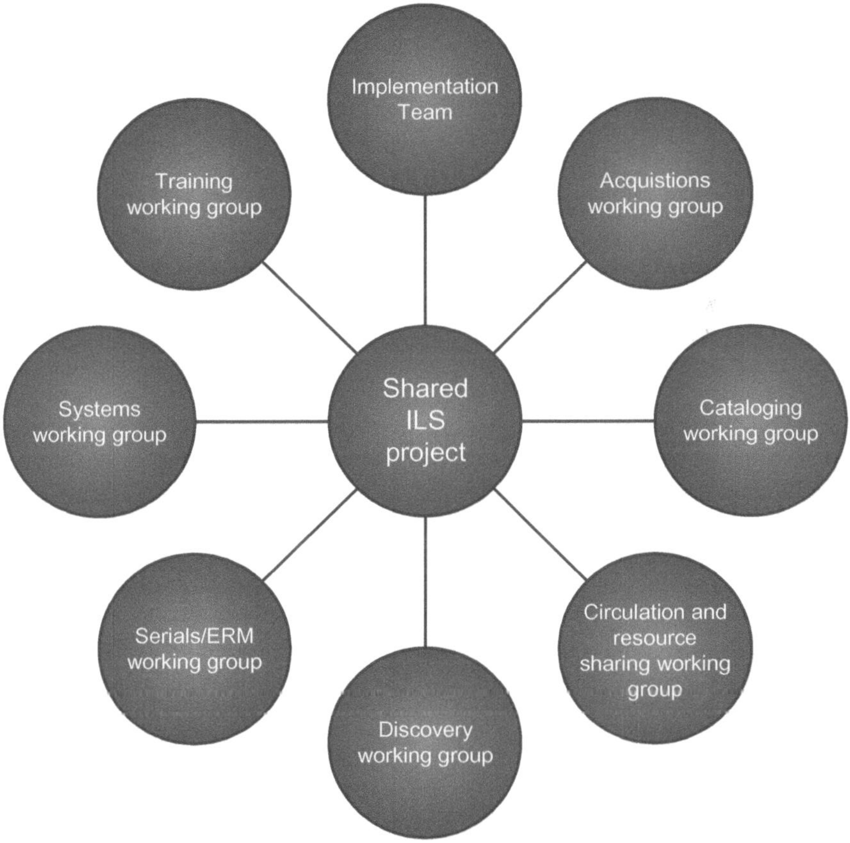

Figure A.1 Orbis-Cascade Alliance Shared ILS project organization.

The original intention of the project was that all libraries would use the Ex Libris product Alma for their staff client and the Ex Libris discovery tool, Primo, as the system front end. This combination proved to be viable for all the libraries in Orbis-Cascade except for the University of Washington. Because it had been a partner with OCLC to develop the Worldcat Local discovery platform, UW had made a large investment in perfecting this discovery tool. When the initial implementation of Primo proved to be less effective than Worldcat Local for search and retrieval, UW decided to retain the OCLC product and offer Primo as a beta product only. After many software upgrades and collaboration with Ex Libris, the University of Washington will be using Primo as its primary discovery tool beginning with the academic year 2015–2016.

Based on the complexities of the shared system project, this transition was an event that challenged everyone involved. Implementing a new system is demanding for any library but doing it while changing workflows and collaborating with new partners (many of them geographically removed from one another) added to the difficulty of the situation. The fear of job loss due to a shared system was high, as staff feared that a gathering momentum for centralized services might result in fewer staff positions on the individual campus libraries.

With the Orbis-Cascade project still in the implementation phase, much of the energy has gone to the mechanics of getting all 37 libraries migrated to the new system. Although difficult, this phase will eventually come to end and the Alliance will then have to grapple with questions that may be even harder to solve to fulfill the Alliance vision. What does it mean for 37 separate institutions to be "one" collection and function as "one" library organization?

The concerns among Alliance members about standards, loss of local autonomy, and patron service customization are very real. Harmonizing all 37 circulation policies, for example, will mean that some libraries may have to lend materials which were formerly for library use only. What will be the reaction of their local patron community when these requested items are no longer on campus but have been loaned out to an Alliance partner? On the other hand, how can the Alliance meet the goal of a single unified collection if some libraries will lend their copy of a resource while other libraries will refuse to lend their copy of the exact same resource?

It is difficult to project at this preliminary stage whether the path being pursued by the Orbis-Cascade Alliance is one that will appeal to many libraries. But it does represent a new vision for library services in a collaborative networked environment. If the Alliance can combine efficient standardized workflows with staff cost savings, it may prove to be a successful model for other libraries to follow.

References

Becker, S. (2009). *RFP matrix*. Prepared for the University of Washington Information School Class LIS 587.

Casey, M. (2005). *Working towards a definition of Library 2.0*. Retrieved from: <http://www.librarycrunch.com/2005/10/working_towards_a_definition_o.html>.

Casey, M., & Savastinuk, L. (2007). *Library 2.0: The librarian's guide to participatory library service*. Medford, NJ: Information Today.

Cohn, J., Kelsey, A., & Fiels, K. (2001). *Planning for integrated systems and technologies: A how-to-do-it manual for librarians*. New York, NY: Neal-Schuman.

Cornish, A., Jost, R., & Arch, X. (2013). Selecting a shared 21st century management system. *Collaborative Librarianship*, *5*(1), 16–28.

Decision Support Tools. (2012). Retrieved from: FOSS4LIB web site <https://foss4lib.org/>.

Griffiths, J., & King, D. (2011). The future of librarians in the workforce: A U.S. perspective. In D. Baker & W. Evans (Eds.), *Libraries and society role, responsibility and future in an age of change*. Witney, Oxford: Chandos Publishing.

Hultman, K. (1998). *Making change irresistible: Overcoming resistance to change in your organization*. Palo Alto, CA: Davies-Black Pub.

Janes, J. (Ed.), (2013). *Library 2020: Today's leading visionaries describe tomorrow's library*. Lanham, MD: Scarecrow Press.

Lerner, F. (2009). *Story of libraries: From the invention of writing to the computer age* (2nd ed.). London: Continuum.

Nix, L. (2010). The library history buff. Library catalog cards. Retrieved from <http://www.libraryhistorybuff.org/catalog-cards.htm>.

O'Reilly, T. (2005). What is Web 2.0: Design patterns and business models for the next generation of software. Retrieved from: <http://oreilly.com/web2/archive/what-is-web-20.html>.

Request for Proposal Shared Library Management Service, (2012). Retrieved from Orbis-Cascade web site <https://www.orbiscascade.org/rfp-for-shared-library-management/>.

Strickland, J. (2008). Is there a Web 1.0? Retrieved from: <http://computer.howstuffworks.com/web-10.htm>.

Toffler, A. (1970). Introduction. In *Future shock*. New York, NY: Random House.

Wilson, T. (1998). *The systems librarian designing roles, defining skills*. Chicago, IL: American Library Association.

Yang, S., & Wagner, K. (2010). Evaluating and comparing discovery tools: How close are we towards next generation catalog? *Library Hi Tech*, *28*(4), 690–709.

Index

FSC
www.fsc.org
MIX
Papier | Fördert gute Waldnutzung
FSC® C083411

Kolibri 360 GmbH
Ferdinand-Jühlke-Straße 7
99095 Erfurt, Deutschland
produktsicherheit@kolibri360.de